SPACE-TIME STRUCTURE

SPACE-TIME STRUCTURE

ERWIN SCHRÖDINGER

PUBLISHED BY THE PRESS SYNDICATE OF THE UNIVERSITY OF CAMBRIDGE
The Pitt Building, Trumpington Street, Cambridge CB2 1RP, United Kingdom

CAMBRIDGE UNIVERSITY PRESS
The Edinburgh Building, Cambridge CB2 2RU, United Kingdom
40 West 20th Street, New York, NY 10011–4211, USA
10 Stamford Road, Oakleigh, Melbourne 3166, Australia

First published 1950
Reprinted 1954 (with corrections), 1960
Reissued as a paperback 1985
Reprinted 1986, 1988, 1991, 1994, 1997

A catalogue record for this book is available from the British Library

ISBN 0 521 31520 4 paperback

Transferred to digital printing 2002

To

MY FRIEND

LOUIS WERNER

IN GRATITUDE

FOR HIS INESTIMABLE AID

CONTENTS

INTRODUCTION

In Einstein's theory of gravitation matter and its dynamical interaction are based on the notion of an intrinsic geometric structure of the space-time continuum. The ideal aspiration, the ultimate aim, of the theory is not more and not less than this: A four-dimensional continuum endowed with a certain intrinsic geometric structure, a structure that is subject to certain inherent purely geometrical laws, is to be an adequate model or picture of the 'real world around us in space and time' with all that it contains and including its total behaviour, the display of all events going on in it.

Indeed the conception Einstein put forward in 1915 embraced from the outset (and not only by the numerous subsequent attempts to generalize it) every kind of dynamical interaction, not just gravitation only. That the latter is usually in the foreground of our mind—that we usually call the theory of 1915 a theory of gravitation—is due to two facts. First, its early great successes, the new phenomena it predicted correctly, were deemed to refer essentially to gravitation, though that is, strictly speaking, true only for the precession of the perihelion of Mercury. The deflexion of light rays that pass near the sun is not a purely gravitational phenomenon, it is due to the fact that an electromagnetic field possesses energy and momentum, hence also mass. And also the displacement of spectral lines on the sun and on very dense stars ('white dwarfs') is obviously an interplay between electromagnetic phenomena and gravitation.

At any rate the very foundation of the theory, viz. the basic principle of equivalence of acceleration and a gravitational field, clearly means that there is no room for any kind of 'force' to produce acceleration save gravitation, which however is not to be regarded as a force but resides on the geometry of space-time. Thus in fact, though not always in the wording, the mystic concept of force is wholly abandoned. Any 'agent' whatsoever, producing ostensible accelerations, does so quâ amounting to an energy-momentum tensor and via the gravitational field connected with the latter. The case of 'pure gravitational interaction' is distinguished only by being the simplest of its kind, inasmuch as the energy-momentum- (or matter-) tensor can here be regarded as located in minute specks of matter (the particles or mass-points) and as having a particularly simple form, while, for example, an electrically charged particle is connected with a matter-tensor spread throughout the space around it and of a rather complicated form even when the particle is at rest. This has, of course, the consequence that in such a case we are in patent need of field-laws for the matter-tensor (e.g. for the electromagnetic field), laws that one would also like to conceive as purely geometrical restrictions on the structure of space-time. These laws the theory of 1915 does not yield, except in the simple case of purely gravitational interaction. Here the defect can at least be camouflaged or provisionally supplemented by simple additional assumptions such as: the particle shall keep together, there shall be no negative mass, etc. But in other cases, such as electromagnetism, a further development of the geometrical conceptions about space-time is called for, to yield the field-laws of the matter-tensor in a natural fashion. This was the second reason for looking upon the theory of 1915 as referring to pure gravitation only.

The geometric structure of the space-time model envisaged in the 1915 theory is embodied in the following two principles:

(i) equivalence of all four-dimensional systems of coordinates obtained from any one of them by arbitrary (point-) transformation;

(ii) the continuum has a metrical connexion impressed on it: that is, at every point a certain quadratic form of the coordinate-differentials,

$$g_{ik}\,dx_i\,dx_k,$$

called the 'square of the interval' between the two points in question, has a fundamental meaning, invariant in the aforesaid transformations.

These two principles are of very different standing. The first, the principle of general invariance, incarnates the idea of General Relativity. I will not commit myself to calling it unshakable. One has occasionally tried to generalize it, and it is difficult to say whether quantum physics might not at some time seriously dictate its generalization. However, the principle as it stands appears to be simpler than any generalization we might contemplate, and there seems to be no reason to depart from it at the outset.

On the other hand, to adopt a metrical connexion straight away does not seem to be the simplest way of getting at it eventually, even if nothing more were intended than an exposition of the 1915 theory. The reason is that the conceptions on which this theory hinges (as invariant differentiation, Riemann-Christoffel-tensor, curvature, variational principles, etc.) are not at all peculiar to the metrical connexion. They come in in a much simpler, more natural and surveyable fashion when you first only introduce as much of a connexion as, and precisely that kind of connexion for which, the idea of 'differentiation' calls out peremptorily in view of the general invariance you have admitted. That is the so-called *affine* connexion. It is then easy, if desired, to specialize it so as to engender a metric.

An important group of attempts to generalize the 1915 theory (inaugurated by H. Weyl as early as 1918) is based on this more general type of connexion.

We shall therefore investigate the geometry of our continuum in three steps or stages, viz.

(1) when only general invariance is imposed;

(2) when in addition an affine connexion is imposed;

(3) when this is specialized to carry a metric.

And we shall find it useful to keep account of which notions are peculiar to each stage, I mean to say which *are* accessible and meaningful at that stage without our having to go to the next one, but have *no* meaning in the previous one.

Many of the statements and propositions worked out in the following apply to any number n of dimensions. But since we are not dealing with pure mathematics but only intend to show the simplest access to possible geometrical models of space-time, we have at the back of our mind always the case $n = 4$. It would be tedious to repeat again and again: this theorem applies to any number of dimensions. Of more interest and importance is the case when a theorem *is* restricted to $n = 4$; therefore this fact will usually be stressed explicitly.

PART I

THE UNCONNECTED MANIFOLD

CHAPTER I

INVARIANCE; VECTORS AND TENSORS

We envisage a (four-dimensional) continuum whose points are distinguished from each other by allotting a quadruplet of continuous labels x_1, x_2, x_3, x_4 to each of them. However, this first labelling shall have no prerogative over any other one

$$\left.\begin{aligned} x'_1 &= x'_1(x_1 \dots x_4), & x'_2 &= x'_2(x_1 \dots x_4), \\ x'_3 &= x'_3(x_1 \dots x_4), & x'_4 &= x'_4(x_1 \dots x_4), \end{aligned}\right\} \tag{1.1}$$

where the x'_k are four continuous, differentiable functions of the x_k, such that their functional determinant vanishes nowhere.†

But, of course, if such a transformation is made, it must be announced and the functions must be indicated, lest the labelling go to the dogs and the points be 'lost'.

Now we are looking out for mathematical entities, numbers or sets of numbers to which a meaning can be attached in such a manifold.

The numerical values of the coordinates are not of that kind, since they change on transformation, and so would any given mathematical function of them, e.g. the sum of their squares. But on the other hand, if there shall be any meaning in safeguarding the *individuality* of every point even on transformation, we must allow that attached to a point may be some *property* that remains, of course, unchanged on transformation. For unless we intend to enunciate some fact concerning that particular point of space-time, what would be the good of labelling it carefully so as to find it again in any frame? Our list of labels would amount to a list of (grammatical) subjects without predicates; or to writing out an elaborate list of addresses without any intention ever to bother who or what is to be found at these addresses.

In the simplest case such a property will be expressed by one number, attached to the point and, by definition, not changing on

† This is necessary in order to secure a one-to-one correspondence between the two sets of labels. But it is well known that exceptions are quite often put up with, as, for example, in the transition from Cartesian to polar coordinates.

transformation. In the way of *illustration* you may think, for example, of the temperature at a given point of a body at a given time. A property expressed by a number that 'by order' is not to be changed on transformation of the frame is called an *invariant* or a *scalar*. We speak of an invariant *field* or scalar *field*, if not only to one particular point but to every point within a certain region a number is attached, all these numbers referring to the same invariant property. Thus a scalar field will be given by a function of the coordinates

$$\phi(x_1, x_2, x_3, x_4),$$

but not by a definite mathematical function. After the transformation (1.1) the same field will be described by substituting for the x_k their values (functions) obtained from the equations (1.1) by solving them; thus if we call these solutions $x_k(x'_1, x'_2, x'_3, x'_4)$, the field will now, in the new frame, be given by

$$\phi[x_1(x'_1, x'_2, x'_3, x'_4), \quad x_2(x'_1 \dots x'_4), \quad x_3(x'_1 \dots x'_4), \quad x_4(x'_1 \dots x'_4)];$$

and this is, of course, an entirely different function of the x'_k from what ϕ was of the x_k. Strictly speaking, we should indicate it by a different letter, say $\psi(x_1, x_2, x_3, x_4)$. The physicist, however, has taken to regarding a definite letter (ϕ in our case) as referring to a *particular* field in *any* frame. His most important general considerations usually refer to 'the general frame', which he does not specialize and therefore has not actually to change very often, though the principle of invariance on transformation is continually at the back of his mind. Whenever he has to contemplate two or more frames simultaneously, say $x_k, x'_k, x''_k, \dots$, he would choose for the functions describing the same scalar field in these various frames the letters

$$\phi, \phi', \phi'', \dots,$$

so that, for example, in the notation used above,

$$\phi[x_1(x'_1 \dots x'_4), \quad x_2(x'_1 \dots x'_4), \quad x_3(x'_1 \dots x'_4), \quad x_4(x'_1 \dots x'_4)] \equiv \phi'(x'_1, x'_2, x'_3, x'_4).$$

For brevity we shall in future write $\phi(x_k)$ instead of $\phi(x_1, x_2, x_3, x_4)$, if it is at all necessary to indicate the arguments. Usually they can be inferred. Also, the dash in ϕ' would indicate that we mean the field-function expressed in the x'_k-frame, without it being necessary to write $\phi'(x'_k)$.

Given (in one frame) two points, P, with coordinates x_k, and $\bar{P}$, with coordinates $\bar{x}_k$, the difference

$$\phi(\bar{x}_k) - \phi(x_k)$$

is also invariant on transformation. Hence also (taking $\bar{P}$ infinitesimally near to P)

$$\frac{\partial \phi}{\partial x_k} dx_k = \text{invariant}. \tag{1.2}$$

(Throughout these lectures we use the convention that the sum from 1 to 4 is to be understood, whenever the same index appears twice in a product.) Indeed since, on transformation,

$$\frac{\partial \phi}{\partial x'_k} = \frac{\partial \phi}{\partial x_l} \frac{\partial x_l}{\partial x'_k} \tag{1.3}$$

and

$$dx'_k = \frac{\partial x'_k}{\partial x_m} dx_m, \tag{1.4}$$

we get

$$\frac{\partial \phi}{\partial x'_k} dx'_k = \frac{\partial \phi}{\partial x_l} \frac{\partial x_l}{\partial x'_k} \frac{\partial x'_k}{\partial x_m} dx_m = \frac{\partial \phi}{\partial x_l} dx_l.$$

The latter (which proves the statement (1.2)) is obtained by summing over k, since $\frac{\partial x_l}{\partial x'_k} \frac{\partial x'_k}{\partial x_m}$ is the partial derivative of x_l (regarded as a function of the undashed x's) with respect to x_m. And that is 1 or 0 according to whether l is the same index as m or different from it.

The array of the four quantities $\partial\phi/\partial x_k$ is itself a mathematical entity with a definite meaning, provided you subject it to the transformation rule (1.3), just as the scalar ϕ was subject to being *not* transformed but simply 'substituted' (German: *umgerechnet*). The meaning of $\partial\phi/\partial x_k$ is that, in *any* frame, it gives you the increment of ϕ (on proceeding to a neighbouring point) as the sum of products, indicated in (1.2), the increments of the coordinates to be taken, of course, in that frame. The entity described by these four partial derivatives is called the gradient of ϕ and is the first example of a property referring to a definite point and given not by one number only, as a scalar is, but by an array of numbers, four in this case. It is the prototype of a *covariant vector*. More especially, it is a covariant vector *field*.

The general conception of a covariant vector is an array of four quantities A_k which 'by order' is to be transformed according to (1.3), thus:

$$A'_k = \frac{\partial x_l}{\partial x'_k} A_l. \tag{1.5}$$

The nature of the entity may (as in the case of the gradient) be such that there is a quadruplet of numbers attached to every point, varying from point to point. Then we speak of a *field*. Or the particular vector might refer just to *one* point. But at all events every vector must refer to *one definite* point, otherwise the prescription (1.5) would be meaningless, we should not know what coefficients to use in it. (What has just been said will refer in the same way and for the same reasons to all the other vectors and tensors to be introduced presently.)

The way in which, according to (1.4) the *differentials of the coordinates* transform is a sort of counterpart to (1.3). We define a *contravariant vector* as an array of four quantities B^k which transform in the same fashion as the dx_k:

$$B'^k = \frac{\partial x'_k}{\partial x_m} B^m. \tag{1.6}$$

By general convention the writing of the index as subscript or as superscript respectively serves to distinguish the 'covariant' and 'contravariant' behaviour. The dx_k themselves are thus an (infinitesimal) contravariant vector, indeed its prototype. With regard to our convention some people write x^k instead of x_k for the coordinates. I do not think this makes for consistency since (i) the x_k themselves are no vector at all and (ii) the symbols $\partial/\partial x_k$ can in many respects be regarded as a (symbolic) *co*variant entity. So it is better to remember that in all these cases the position of the *whole differential* (whether it stands in the numerator or in the denominator) replaces, as it were, the position of the *index*.

From (1.5) and (1.6) follows immediately

$$A'_k B'^k = A_k B^k = \text{invariant}. \tag{1.7}$$

It is called the inner or scalar product. When it is zero, the two vectors are by some people called pseudo-orthogonal.

Given several (s) vectors at the same point, partly covariant, partly contravariant, the array of 4^s quantities

$$A^k B^l C^m \ldots G_p H_q \ldots \tag{1.8}$$

follows a linear transformation law that can easily be made out from (1.5) and (1.6), but we need not write it out explicitly. An array of 4^s quantities which follow *this* transformation law is called a tensor of rank s and indicated by a symbol like

$$T^{klm\ldots}{}_{pq\ldots}, \tag{1.9}$$

where, of course, the number of superscripts and the number of subscripts must be given separately, fully to characterize the nature of the entity T. The product (1.8) is a special case of such a tensor, but not the most general tensor of this kind, since it depends only on $4s$ independent numbers and

$$4s < 4^s$$

for $s > 1$. The order of the superscripts in the notation (1.9) is relevant, indeed $T^{lkm\ldots}{}_{pq\ldots}$ would in the particular case (1.8) mean $A^l B^k C^m \ldots G_p H_q \ldots$, which is different from (1.8).

It is not the same tensor, but it is a tensor of the same type. It is worth showing that it really has *exactly* the same transformation rule. An example will suffice. Take a contravariant tensor of the third rank T^{klm}. It transforms thus:

$$T'^{klm} = \frac{\partial x'_k}{\partial x_r} \frac{\partial x'_l}{\partial x_s} \frac{\partial x'_m}{\partial x_t} T^{rst}.$$

Exchange k and l and at the same time the notation of the summation indices r, s

$$T'^{lkm} = \frac{\partial x'_l}{\partial x_s} \frac{\partial x'_k}{\partial x_r} \frac{\partial x'_m}{\partial x_t} T^{srt}.$$

The coefficient is unchanged, but in the T-arrays the first two indices have been exchanged. The point is that you may regard the T^{123} component as the (213)-component etc. of *another* tensor. The same would hold for any permutation, provided you make the same permutation in *all* components.

The same holds, of course, for subscripts. But at the moment there is no relevant order between subscripts and superscripts.

The two types of *vectors* are clearly special cases of tensors, viz. the tensors of rank 1. A scalar may be called a tensor of rank zero.

By multiplying the components of any two tensors in all combinations:

$$T^{klm\ldots}{}_{pq\ldots} S^{abc\ldots}{}_{rst\ldots}$$

you get again a tensor. That is clear from the transformation rules. It is called the outer or direct product of the two.

If in (1.9) you execute a summation with respect to an upper and a lower index, as for example,

$$T^{klm\ldots}{}_{kq\ldots}, \tag{1.10}$$

it is again easy to show from the transformation rule (which we have indicated, but not written out) that this is a tensor with rank two less than the original one. It could be indicated by a symbol like

$$S^{lm\ldots}{}_{q\ldots}. \tag{1.11}$$

This process of forming from a given tensor which has at least one index of each kind a tensor of lower rank is called contraction (German: *Verjüngung*). Observe that (1.9) admits of various contractions. The tensor, e.g.

$$T^{klm\ldots}{}_{pk\ldots}, \tag{1.12}$$

is distinctively different from (1.10), though of the same general type, i.e. the same rank and the same number of superscripts and subscripts.

Tensors can be added or subtracted or, more generally, linearly combined with either constant or invariant (scalar) coefficients, if and only if they are of exactly the same type and refer to exactly the same point of the continuum. By 'can be' we mean that in this and only in this case, the result will again have a simple transformation formula, to wit it is a tensor of the same type and referring to the same point.

The most important number in mathematics is the zero. Our present sign for it as well as the word zero comes from the Arabs. (It is, by the way etymologically the same as English cipher, French *chiffre*, German *Ziffer*, which have, however, acquired a different meaning.) But the notion is older, it turns up in Babylonian Mathematics soon after 1000 B.C.† and may have been received from India. Let me dwell for a moment on the importance of this concept. A great many of our propositions and statements in mathematics take the form of an equation. The essential enunciation of an equation is always this: that a certain number is zero. Zero is the only number with a charter, a sort of royal privilege.

† V. Gordon Childe, *Man Makes Himself* (London: Watts and Co., 1936), pp. 222 and 255.

While with any other number any of the elementary operations may be executed, it is prohibited to *divide* by zero—just as, for example, in many houses of parliament *any* subject may be discussed, only the person of the sovereign is excluded. If you divide by zero, nonsense is usually the result. This prerogative is essential, you have to think of it every minute; whenever you divide, you must satisfy yourself that the divisor is not 'of royal blood', that it is *not* zero. Another consequence is that royal blood cannot (by multiplication) be obtained otherwise than from royal blood. A product cannot vanish unless at least one of its factors vanish. It is not accidental that more often than not the conclusion of a proof runs thus: $AB = 0$, $B \neq 0$, $\therefore A = 0$.

In the same way the most important tensor of any type is the zero tensor of that type, that is, the one whose components all vanish. It is a *numerically invariant tensor*, since the transformation formulae are linear and homogeneous. That is the reason why tensors play the all-important part they play. For it has the consequence that an equation of the following kind between two tensors S and T

$$S^{kl\ldots}{}_{pq\ldots} = T^{kl\ldots}{}_{pq\ldots}$$

is independent of the frame (for it means that $S^{\cdots}{}_{\cdots} - T^{\cdots}{}_{\cdots}$ is the zero tensor)—provided, of course, that S and T are of the same type and refer to the same point. If they did not, this would not hold, the above equation would be meaningless, and therefore we shall never contemplate that sort of thing.

Perhaps this is the place to mention a convention, which is always made *tacitly*, though it would deserve to be mentioned explicitly, just as the 'summation convention', of which it is the counterpart. According to the latter an index that appears twice in the same product is to include summation from 1 to 4. Now an index that appears only once, but then, of course, in *every term* of an equation, implies that the equation holds for any value 1 to 4 of that index. By the first convention we shove many terms of an equation into *one*, by the second we shove many equations into one. For example, if you write

$$S^{klm}{}_{m} = R^{kl},$$

this represents in general 16 equations, each of which has four terms on the left.

An important application of the invariance of tensor equations is to the symmetry of tensors. If for a tensor S one of the following two equations

$$S^{kl\ldots}{}_{pq\ldots} = \pm S^{lk\ldots}{}_{pq\ldots}$$

holds in one frame, it holds in every frame. We then call S symmetric or antisymmetric, respectively, with respect to its first pair of superscripts. The same could happen for the pair p and q, but not, for example, for the pair k and q. (Symmetry might happen to obtain in one particular frame, but it would be of no interest, just a chance event). Later we shall come to know more complicated symmetry properties. As a corollary we note that a general tensor can always be decomposed into a sum of two tensors, one of which is symmetric, the other one skew with respect to a certain pair of indices of the same character. Similar theorems hold also for more complicated forms of symmetry.

Given a tensor with t contravariant and r covariant indices, contemplate any t covariant and r contravariant vectors and form the contracted product

$$S^{kl\ldots}{}_{pq\ldots} A_k B_l \ldots F^p G^q \ldots. \tag{1.13}$$

Then from the rules for outer and inner multiplication, this product (which is just *one* number, all the indices being 'killed' by contraction) is an invariant.

It is interesting and useful to know that the converse is also true: if you know nothing about the array of numbers $S^{\cdots}{}_{\cdots}$ but that the 'product' (1.13) is an invariant for *any* set of vectors $A \ldots G \ldots$, then the $S^{\cdots}{}_{\cdots}$ are the components of a tensor of the type defined by its indices. This inverse theorem (which we shall prove forthwith) might serve as an alternative definition of a tensor; but more important is that it is frequently used to establish the tensor-property of an array of numbers, for which it is not yet secured.

To prove this inverse theorem, envisage a particular transformation, call $S'^{\cdots}{}_{\cdots}$ the components of S transformed *as if* $S^{\cdots}{}_{\cdots}$ were a tensor, and $S''^{\cdots}{}_{\cdots}$ any set of numbers sharing with the $S'^{\cdots}{}_{\cdots}$ the property that they make (1.13) invariant on this particular transformation for any set of vectors $A \ldots G \ldots$. By subtracting the two equations which express that both $S'^{\cdots}{}_{\cdots}$ and $S''^{\cdots}{}_{\cdots}$ make (1.13) an invariant, you get

$$(S'^{kl\ldots}{}_{pq\ldots} - S''^{kl\ldots}{}_{pq\ldots}) A'_k B'_l \ldots F'^p G'^q \ldots = 0.$$

Now since the original components of the vector $A \ldots G$ were quite arbitrary, the same holds for the primed components, since the transformation-formulae (1.5) and (1.6) have non-vanishing determinants. Hence you can choose the vectors so that of A' only the kth, of B' only the lth ... of G' only the qth component is different from zero. Then you get

$$S'^{kl\ldots}{}_{pq\ldots} - S''^{kl\ldots}{}_{pq\ldots} = 0,$$

saying that these particular two numbers of the arrays S' and S'' are equal. Obviously by suitably different choices of the vectors the same can be shown for any pair out of the S' and S'' and thus our assertion is proved.

Simple corollaries of our theorem are illustrated by the following example. If we know that

$$S^{kl}A_l = \text{contravariant vector}$$

for any choice of the covariant vector A, then S^{kl} is a contravariant tensor of second rank. Naturally. For if the above is a contravariant vector for any choice of the vector A, then

$$S^{kl}A_l B_k = \text{invariant}$$

for any choice of the vectors A and B.

As can be seen from our proof, it is vital that the invariance of the product be warranted for *arbitrary* vectors. However, a certain remission can be granted if something more is known about the array of the S. To give an example, if it is only warranted that

$$S^{kl} A_k A_l = \text{invariant}$$

for any choice of the vector A, but if in addition it is known that in any frame

$$S^{kl} = S^{lk}$$

(symmetry), the tensor property of S can be proved along the lines followed above. (Without the symmetry one could only show that $S^{kl} + S^{lk}$ is a tensor.)

As an example of the general method we prove the tensor property of the mixed unity tensor, which in itself is an important entity. Envisage the array of 16 numbers

$$\delta^i_k,$$

with numerical value 0 or 1 according to whether $i \neq k$ or $i = k$.

Then for any pair of vectors at any point of the continuum

$$\delta^i_k A_i B^k = A_k B^k = \text{invariant},$$

according to (1.7). Hence δ^i_k is a mixed tensor and is correctly written with one superscript and one subscript. It is one of the (very few) *numerically* invariant tensorial entities, that is to say even its components are the same in every frame. One feels tempted to call it a symmetrical tensor. However, this would not be appropriate. For symmetry with respect to two indices of different character is in general not preserved on transformation. That it is so here is an exceptional occurrence.

Notice, by the way, that even the more trivial statement, that

$$\delta^i_k B_i = B_k$$

is a vector for any B_i, would suffice to infer the tensor-property of δ^i_k.

CHAPTER II

INTEGRALS. DENSITIES. DERIVATIVES

INTEGRALS. DENSITIES

The subject-matter of the previous chapter is called tensor *algebra*. It is characterized by the fact that only relations between invariants, vectors or tensors referring to the same point of the continuum are contemplated. From the point of view taken here,† algebraic relations between vectors and tensors referring to different points are meaningless.

Remember, however, that we based the notion of tensors on that of vectors, and the latter on the notion of the gradient, and there is hardly any simple and natural alternative to this procedure. Now in forming the gradient we actually had to compare the values of an *invariant* at different points, and at the same time we made the first step at introducing *analysis* into our continuum. In this and the following chapters we shall have to extend it. Analysis will involve derivatives and integrals. We shall have to study both from the point of view of general invariance. However, this does not mean to look out only for invariants, but also for entities with tensorial character, because, as we have seen, an equation between them (or in other words a system of equations saying that a tensor vanishes) is conserved on transformation. We begin with space-time-*integrals*. That leads to a certain extension of the notion of tensors, viz. to tensor densities.

We had emphasized that there is no point in adding (or, more generally, in forming linear aggregates of) tensors or vectors referring to different points. This would have no simple meaning. For instance, an equation stating that a vector A at a point P equals a vector B at a different point Q, even if it happens to obtain in one frame, is entirely uninteresting, because it is destroyed by trans-

† Only quite recently an attempt has been made to envisage a connexion that involves *algebraic* relations between tensors at different points. See A. Einstein and V. Bargmann, *Ann. Math.* XLV, pp. 1 and 15, 1944. See also E. Schrödinger and F. Mautner, *Proc. R. Irish Acad.* L, 143 and 223, 1945. These attempts are *not* included in the present exposition.

formation. Or again, let A^k be a contravariant vector field and contemplate the four integrals

$$\iiiint A^k dx_1\,dx_2\,dx_3\,dx_4,$$

taken over a given region of space-time, and, of course, over the exactly *corresponding* region in any other frame. (Integrals of this type will in future be abbreviated thus: $\int A^k dx^4$.) Now the above integrals are neither invariants nor are they the components of a contravariant vector—they are devoid of sense and interest.

But if A were an *invariant* (scalar) and we formed in the same way

$$\int A dx^4$$

(always over an invariantly fixed domain), would that be an invariant? Obviously not. Though there is no objection to adding *invariants* that refer to different points, yet we know that on transforming

$$\int A dx^4 = \int A \left|\frac{\partial x_k}{\partial x'_i}\right| dx'^4,$$

thus

$$\neq \int A dx'^4$$

in general.† For the equation

$$\int A dx^4 = \int A' dx'^4$$

to hold, in other words for the integral to be an invariant, the 'transformation law' for A would have to be not

$$A' = A,$$

but

$$A' = \left|\frac{\partial x_k}{\partial x'_i}\right| A,$$

that is, it would by definition have to take on as a factor the functional determinant appearing in the transformed integral in consequence of the transformation of the 'product of the differentials'.

We give a quantity behaving in that way the name of *scalar density*. It has become customary to denote a density by a Gothic

† To make the integral invariant, one would have to restrict the allowed transformations by the condition that their functional determinants be 1. This would be inconvenient.

letter. It will prove convenient to extend the notion of 'density' to more-component entities† which bear to tensors the same relation as the scalar density to a scalar, namely just to have their transformation formulae enhanced by a factor, the determinant $|\partial x_k/\partial x'_i|$ —always this one, irrespective of the character of the other indices. To make the point quite clear let us write out *in extenso* the transformation formula for a general tensor-density

$$\mathfrak{A}^{kl\ldots}{}_{pq\ldots}. \tag{2.1}$$

It reads

$$\mathfrak{A}'^{kl\ldots}{}_{pq\ldots} = \left|\frac{\partial x_j}{\partial x'_i}\right| \frac{\partial x'_k}{\partial x_m}\frac{\partial x'_l}{\partial x_n}\cdots\frac{\partial x_r}{\partial x'_p}\frac{\partial x_t}{\partial x'_q}\cdots\mathfrak{A}^{mn\ldots}{}_{rt\ldots}.$$

Densities obviously share with ordinary tensors the property that they (i.e. all their components) vanish in every frame, when they do so in one frame. For this vital property resided only on the homogeneous linear character of the transformation. Hence they are equally useful; equations between such as refer to the same point are independent of the frame, they persist on transformation.

In order to get hold of a scalar or tensor density, we need not snatch it from the sky; such entities can be constructed from the tensors we have introduced previously.

Envisage a covariant antisymmetric tensor of the fourth rank

$$T_{klmn}.$$

By antisymmetric we mean that an exchange of *any* two subscripts should just merely produce a change of sign of the component. If we denote the numerical value of T_{1234} by a capital Gothic $\mathfrak{T}$ (why Gothic, will appear forthwith) any other component T_{klmn} is thus $\pm\mathfrak{T}$, according to whether the permutation *klmn* is odd or even, while components with not all their subscripts different from one another vanish, of course. Now write out the transformation formula for the component T_{1234}

$$T'_{1234} = \frac{\partial x_k}{\partial x'_1}\frac{\partial x_l}{\partial x'_2}\frac{\partial x_m}{\partial x'_3}\frac{\partial x_n}{\partial x'_4}T_{klmn}.$$

Considering the values of the T_{klmn}, this gives

$$T'_{1234} = \left|\frac{\partial x_k}{\partial x'_i}\right| T_{1234} = \left|\frac{\partial x_k}{\partial x'_i}\right| \mathfrak{T}.$$

† Do not infer, please, that the integral of the component of a tensor density (other than scalar density) *has* a meaning! *It has not.*

Or, if we use the consistent notation $\mathfrak{T}'$ for T'_{1234}, then

$$\mathfrak{T}' = \left|\frac{\partial x_k}{\partial x'_i}\right| \mathfrak{T}.$$

Thus an alternative way of looking upon our covariant antisymmetric tensor of the fourth rank is to regard it as an entity with only one component, though not as a scalar, but as a scalar density.

The theorem can be sort of reversed. Let A be a scalar. Envisage an entity $\mathfrak{E}^{klmn}$ (why we choose a Gothic letter will appear forthwith) defined in any frame by $\mathfrak{E}^{klmn}$ being $\pm A$ according to the sign of the permutation ($klmn$), but zero if the four superscripts are not all different. A queer but correct way of expressing that A is an invariant ($A' = A$) is then

$$\mathfrak{E}'^{klmn} = \left|\frac{\partial x_i}{\partial x'_j}\right| \frac{\partial x'_k}{\partial x_r}\frac{\partial x'_l}{\partial x_s}\frac{\partial x'_m}{\partial x_t}\frac{\partial x'_n}{\partial x_u}\mathfrak{E}^{rstu}.$$

Indeed, the prescribed summations yield a functional determinant which just cancels the one in front and we are left with $\mathfrak{E}'^{klmn} = \mathfrak{E}^{klmn}$. But this 'queer but correct' formula tells us that $\mathfrak{E}$ is a contravariant antisymmetric tensor density of rank 4. It is customary to denote it, in the particular case $A = 1$, by

$$\epsilon^{klmn}.$$

This ϵ-density is a valuable acquisition, a very often used tool. It is, by the way, a further *numerically* invariant entity that we encounter.

You can, for instance, from ϵ and a covariant antisymmetric tensor of the second rank ϕ_{kl} form the following entity which is, clearly, a scalar density

$$\tfrac{1}{8}\epsilon^{klmn}\phi_{kl}\phi_{mn}, \tag{2.2}$$

that in plain writing reads

$$\phi_{12}\phi_{34} + \phi_{23}\phi_{14} + \phi_{31}\phi_{24}. \tag{2.3}$$

Also

$$\tfrac{1}{2}\epsilon^{klmn}\phi_{kl} = \mathfrak{f}^{mn} \tag{2.4}$$

is a contravariant antisymmetric density of the second rank. In plain words: given an (antisymmetric) tensor ϕ_{kl} you may regard ϕ_{12} as the (34)-component, ϕ_{23} as the (14)-component... ϕ_{34} as the (12)-component of *another* entity, but this other entity is contravariant and not a simple tensor but a density. These simple facts alone, if you consider the big part antisymmetric second-rank tensors play, would suffice to show that it is useful to extend the notion of density to others than just scalar densities.

That (2.2) or (2.3) is a scalar density, forms a special case of a more general theorem, about forming a scalar density from *any* covariant second-rank tensor. Let g_{ik} be such a one, so that, on transformation,

$$g'_{ik} = \frac{\partial x_l}{\partial x'_i}\frac{\partial x_m}{\partial x'_k} g_{lm}. \tag{2.5}$$

Here the right-hand side can, for the moment, be regarded as a 'matrix product' of the matrices $\partial x_l/\partial x'_i$, g_{lm} and $\partial x_m/\partial x'_k$ (in *this* order!) Thus from a well-known theorem about the determinant of a product-matrix you get, if you call g' the determinant of g'_{ik} and g that of g_{ik}:

$$g' = \left|\frac{\partial x_i}{\partial x'_j}\right|^2 g, \tag{2.6}$$

hence

$$\sqrt{g'} = \left|\frac{\partial x_i}{\partial x'_j}\right| \sqrt{g}. \tag{2.7}$$

In words: the square root of the determinant of any covariant second-rank tensor is a scalar density. The case of a symmetrical tensor g_{ik} will be of importance in metrical geometry (Einstein's 1915 theory). In the case of a skew-symmetrical tensor, the square root can be extracted and leads precisely to (2.2) or (2.3), as is easy to verify directly.

We use the occasion to demonstrate another important fact. Take g to be $\neq 0$. The minor of g_{ik} in the determinant g we denote by M^{ik}, without prejudice as to its tensorial character. Then by a well-known theorem about determinants

$$g_{mk} M^{lk} = \delta^l{}_m g. \tag{2.8}$$

This holds, of course, in any frame, thus also for the primed quantities, provided M'^{lk} always means the minors in that frame. But since from (2.8) or, say, from

$$g_{mk}\frac{M^{lk}}{g} = \delta^l{}_m \tag{2.9}$$

the quantities M^{lk}/g are determined uniquely and since (in virtue of the tensor property of $\delta^l{}_m$) the preceding equation will also hold in any frame for *those* quantities which are obtained from the M^{lk}/g by transforming them as a contravariant tensor of the second rank, it follows that they actually form such a tensor. *The 'normalized minors' of any covariant tensor of the second rank form a contravariant tensor of the second rank.* It is easy to prove that in this statement the terms covariant and contravariant can be exchanged. Moreover, if from the tensor

$$\frac{M^{lk}}{g} = g^{lk} \quad \text{(say)} \tag{2.10}$$

you again form the normalized minors, you fall back on the tensor g_{lk}.

If instead of (2.10) you contemplate the array

$$\frac{M^{lk}}{\sqrt{g}} = \mathfrak{g}^{lk} \quad \text{(say)},$$

they form, of course, a contravariant tensor *density* of the second rank.

It is noteworthy that in the case of a skew-symmetric tensor ϕ_{ik} this tensor density is the same as the one arrived at in (2.4) in a different way, as is easy to show by directly computing the minors in this case.

We mentioned above that g is in this case the square of the scalar density (2.3), for which we introduce the notation $\mathfrak{J}_2$:

$$\tfrac{1}{8}\epsilon^{klmn}\phi_{kl}\phi_{mn} \equiv \phi_{12}\phi_{34} + \phi_{23}\phi_{14} + \phi_{31}\phi_{24} = \mathfrak{J}_2. \tag{2.11}$$

Hence by applying (2.9) to this case, we get

$$\mathfrak{f}^{lk}\phi_{mk} = \delta^l{}_m \cdot \mathfrak{J}_2. \tag{2.12}$$

From (2.4) this can also be written

$$\tfrac{1}{2}\epsilon^{hilk}\phi_{hi}\phi_{mk} = \delta^l{}_m \cdot \mathfrak{J}_2. \tag{2.13}$$

By contracting with respect to l and m you fall back on (2.11), since $\delta^m{}_m = 4$. But, of course, (2.12) or (2.13) contain more than (2.11). In matrix language it tells you that the matrix product of the matrices $\mathfrak{f}^{ik}$ and ϕ_{ik} is a multiple of the unity matrix, which cannot be grasped directly from the definition (2.4).

As last examples for constructing densities from tensors, let us first envisage a covariant antisymmetric tensor of the third rank A_{ikl}. Disregarding the sign, it has only four non-vanishing numerically different components, according to which of the four subscripts 1, 2, 3, 4 is absent. Now, with the help of the tensor density ϵ you can form from A the contravariant vector density

$$\tfrac{1}{6}\epsilon^{klmn}A_{klm} = \mathfrak{A}^n \quad \text{(say)}.$$

The correlation is very simple, you can formulate it thus: a covariant antisymmetric tensor of third rank can always be regarded as a contravariant vector density of which the nth component is the klm-component of the tensor, $klmn$ forming an even permutation of 1234.

Vice versa, from a *co*variant vector B_k you can form the antisymmetric contravariant density of the third rank, thus

$$\epsilon^{klmn}B_n = \mathfrak{Q}^{klm},$$

where the first member comprises only one term, because n has to be the fourth index with respect to k, l, m.

Comprehensively, the relationship between totally skew tensors and tensor densities is the following. From the following covariant skew tensors†

$$A, \quad A_i, \quad A_{ik}, \quad A_{ikl}, \quad A_{iklm}$$

contravariant skew densities of *complementary rank* can be derived by multiplying them with the ϵ-density and contracting with respect to all the original subscripts

$$\mathfrak{A}^{iklm}, \quad \mathfrak{A}^{klm}, \quad \mathfrak{A}^{lm}, \quad \mathfrak{A}^{m}, \quad \mathfrak{A}.$$

If the factors $\qquad 1, \quad 1, \quad \tfrac{1}{2}, \quad \tfrac{1}{6}, \quad \tfrac{1}{24}$

are included, the derived density has the same components as the tensor, only in different labelling.

There is *no* corresponding theorem about *co*variant densities and contravariant vectors, simply because we are practically not interested in tensorial entities that take on another than the *first* power of the functional determinant on transformation. (Such a thing as, for example, $\epsilon^{klmn}\mathfrak{A}_{lmn}$ would take on the second power of this determinant.)

For practical purposes it may be useful to record the following rules.

† An invariant may range with co- or contravariant tensors, and both an invariant and a vector may be ranged with skew tensors, provided you define (as you may) a covariant/contravariant tensor as one that has no contravariant/covariant index, and define totally 'skew or antisymmetric' by: changing sign on exchange of an index (if any) with any other one (if there is another one) of the same kind.

Any 'juxtaposition' of tensors is again a tensor, whose nature is to be seen from the total array of upper and lower indices, disregarding such as appear twice, in both positions (summation or dummy indices). Take care never to use a letter twice 'by accident', let alone using it more than twice!

Only entities of exactly the same type can be added or subtracted or put equal. Hence an index must either appear in every term of the equation in the same position or twice in the same term in different positions (summation index).

One (but only one) of the entities 'juxtaposed' in a term is allowed to be a density; then the term is a density and all the terms of the equation must be of this type.

The rule of not using a letter again 'by accident' does not refer to summation indices in different terms. Here no confusion can be caused in this way.

A statement which we might have made earlier about outer products and which, in all its simplicity, is quite important is this. If the product is a purely 'outer' product, i.e. if the juxtaposition involves no further contraction, it can only vanish if at least one of its factors is a zero tensor. In other words, there are no 'divisors of zero' in the algebra of tensors and tensor densities.

DERIVATIVES

For shortness we shall henceforth occasionally indicate the derivative with respect to x_k by a lower index k, preceded by a comma.

Except in the case of an invariant, the derivative of a tensor-component, as for example,

$$A_{k,i}$$

has no proper meaning, because it results from subtracting tensors referring to different points, viz. the A_k in the point x_l from the A_k in a certain neighbouring point. (One must not think that this small shift 'does not matter', for in the derivative we contemplate precisely the change in A_k produced by this small shift.)

If we compute, for example, from

$$\frac{\partial \phi}{\partial x'_k} = \frac{\partial x_l}{\partial x'_k} \frac{\partial \phi}{\partial x_l}$$

the transformation formula for the second derivatives

$$\frac{\partial^2 \phi}{\partial x'_k \partial x'_i} = \frac{\partial x_l}{\partial x'_k}\frac{\partial x_m}{\partial x'_i}\frac{\partial^2 \phi}{\partial x_l \partial x_m} + \frac{\partial^2 x_l}{\partial x'_k \partial x'_i}\frac{\partial \phi}{\partial x_l} \tag{2.14}$$

we see that not only do they not form a tensor, but they do not even share the feature, that their vanishing is an invariant property. The same holds, of course, for any covariant vector field. From its transformation formula

$$A'_k = \frac{\partial x_l}{\partial x'_k} A_l \tag{2.15}$$

you get by differentiation

$$\frac{\partial A'_k}{\partial x'_i} = \frac{\partial x_l}{\partial x'_k}\frac{\partial x_m}{\partial x'_i}\frac{\partial A_l}{\partial x_m} + \frac{\partial^2 x_l}{\partial x'_i \partial x'_k} A_l, \tag{2.16}$$

which is exactly the same as (2.14), only for an arbitrary A_l (not just, as there, a gradient). Again we see that $A_{l,m}$ behaves like a covariant second-rank tensor, except for the additional term, containing the non-differentiated A_l and the second derivatives of the transformation. This again has the effect that our array of derivatives would not necessarily vanish in the primed system, as a consequence of their vanishing in the unprimed.

A very similar state of affairs obtains, as one easily realizes, for any tensor or tensor density.

There are, however, certain linear combinations of derivatives of tensor-components in which the terms containing the second derivatives of the coordinates together with undifferentiated components of the original tensor cancel. These linear combinations are then tensors, the index of derivation always playing the role of a covariant index (subscript). They are easily remembered. They are all completely antisymmetric. We begin with tensors. The first one we know already.†

(1) The gradient of an invariant: $\phi_{,k}$. This is a covariant vector. If from it you form what one calls (new definition!) the *curl*

$$\phi_{,k,i} - \phi_{,i,k} = 0,$$

you get zero. This shows that the additional terms must cancel in this difference, as you can see by direct inspection of (2.14). But

† There is no harm in reckoning a scalar among the antisymmetric tensors! See the footnote on page 20.

we can also see that they must cancel in the curl of *any* covariant vector. Hence:

(2) The curl of a covariant vector A_k: $\frac{\partial A_k}{\partial x_i} - \frac{\partial A_i}{\partial x_k}$ is a covariant antisymmetric tensor of second rank.

Now the game goes on. If you form of it what is called (new definition!) the *cyclical divergence*:

$$(l \neq k \neq i) \qquad \frac{\partial}{\partial x_l}\left(\frac{\partial A_k}{\partial x_i} - \frac{\partial A_i}{\partial x_k}\right) + \text{the two cyclical terms} = 0.$$

Hence, here too, the terms containing the non-differentiated second-rank tensor must cancel. And they must do so for *any* covariant skew tensor of the second rank. Hence

(3) The cyclical divergence of a covariant skew second-rank tensor ϕ_{ik}

$$\frac{\partial \phi_{ik}}{\partial x_l} + \frac{\partial \phi_{kl}}{\partial x_i} + \frac{\partial \phi_{li}}{\partial x_k}$$

is a totally antisymmetric covariant third-rank tensor. In continuing you must be careful. If you formed $\partial/\partial x_m$ of this tensor and added the cyclical permutation it would *not* vanish. You must introduce a $(-)$ sign whenever the permutation is odd. Hence also

(4) From an antisymmetric covariant third-rank tensor A_{ikl} the following sum of four derivatives†

$$\Sigma(-1)^! \frac{\partial}{\partial x_m} A_{ikl}$$

is an antisymmetric tensor of fourth rank.

That is all. You cannot continue, because there are only four indices. (In more dimensions you could.)

Now, on account of the correspondence between antisymmetric tensors and densities, four similar statements about densities follow, I will label them (1′), (2′), (3′), (4′).

(4′) The divergence (new definition!) of a contravariant vector density $\mathfrak{A}^k$ to wit $\partial \mathfrak{A}^k/\partial x_k$ is an invariant density.

(3′) The divergence ('tensor divergence', new definition!) of an antisymmetric contravariant tensor density of second rank $\mathfrak{A}^{kl}$ to wit $\partial \mathfrak{A}^{kl}/\partial x_l$ is a contravariant vector density.

† The symbolic exponent (!) is to remind you of what has just been said about the sign.

(2′) The tensor divergence (new definition! though the same word is used) of a skew third-rank contravariant tensor-density $\mathfrak{A}^{klm}$ to wit $\partial\mathfrak{A}^{klm}/\partial x_m$ is a second-rank density of the same description. And finally

(1′) The tensor divergence (see above bracket) of a skew contravariant fourth-rank tensor density $\mathfrak{A}^{klmn}$ to wit $\partial\mathfrak{A}^{klmn}/\partial x_n$ is a third-rank density of the same description.

To the best of my knowledge these are all the linear aggregates of first derivatives of tensors or tensor densities that have tensor character. The most relevant ones are (1), (2), (3), (4′), (3′).

The *vanishing* of one of the above-derived tensors has in all cases a good meaning, viz. for (1) that the scalar ϕ is constant, (2) that the vector A_k is a gradient. (3) and (3′) are exemplified by Maxwell's vacuum-equations and (4′) indicates (or is usually expressed by saying) that the current $\mathfrak{A}^k$ is source-free.

Yet they are not sufficient to establish an exhaustive tensor analysis in our continuum. Not even such a simple question as this has any meaning: When is a vector field A_k to be considered as *constant* throughout a certain region? For the vanishing of all the derivatives $A_{k,i}$ is (as we have seen) not a frame-independent property, because $A_{k,i}$ is not a tensor.

The geometrical concept for removing this difficulty will be introduced in Part II. Before doing so, let us dwell more closely on the interesting fact alluded to just above, that the analytical means developed up to here suffice to establish the principal statements of Maxwell's theory, which may duly be called the spiritual ancestor of all field theories that were to follow it. The elementary form of Maxwell's equations reads in the familiar notation of three-dimensional vector calculus:

$$\left.\begin{aligned} \operatorname{curl} H - \dot{D} &= I \\ \operatorname{div} D &= \rho \end{aligned}\right\} \qquad \text{(A)}$$

$$\left.\begin{aligned} \operatorname{curl} E + \dot{B} &= 0 \\ \operatorname{div} B &= 0 \end{aligned}\right\} \qquad \text{(B)}$$

(the units to be chosen so as to remove factors 4π or c). The behaviour usually, and naturally, attributed to current and charge (I,ρ) on an elementary change of the scale of length prompts us to regard them as *densities*, and thus to look upon the quadruplet (A) as equations between densities. We must then unite the elementary vector quantities H and D into a contravariant antisymmetric tensor density of the 2nd rank $\mathfrak{f}^{ik}$ in such a way that the

components of H correspond to $\mathfrak{f}^{23}, \mathfrak{f}^{31}, \mathfrak{f}^{12}$;
components of D correspond to $\mathfrak{f}^{41}, \mathfrak{f}^{42}, \mathfrak{f}^{43}$.

Then equations (A) read $$\frac{\partial \mathfrak{f}^{ik}}{\partial x_k} = \mathfrak{s}^i, \qquad \text{(A}'\text{)}$$

where the four-current $\mathfrak{s}^k$ replaces (I, ρ). In the case of the second quadruplet (B) no preference suggests itself as to their character (tensors or densities), indeed the choice is irrelevant. It is only a question of nomenclature, since the ϵ-density allows a simple transition from one to the other. (So it does in the first set. But if there we choose a *covariant tensor* in lieu of $\mathfrak{f}^{ik}$, we have to take a covariant antisymmetric tensor of the third rank in lieu of $\mathfrak{s}^k$; as Einstein has once suggested, and for very good reasons.)

Keeping to the usual nomenclature, we unite E, B to a covariant skew tensor ϕ_{ik}, in such a way that the

components of B correspond to $\phi_{23}, \phi_{31}, \phi_{12}$;
components of E correspond to $\phi_{14}, \phi_{24}, \phi_{34}$.

Then equations (B) read

$$\frac{\partial \phi_{ik}}{\partial x_l} + \text{two cyclical terms} = 0. \qquad \text{(B}'\text{)}$$

By (A′) and (B′) we have established Maxwell's fundamental equations invariantly in an arbitrary frame, using nothing but the means developed hitherto in these lectures; that is, for an *unconnected* space-time-manifold (neither affinity nor metric has been introduced). What we *cannot* establish in this manner is the relationship between the density (H, D) or $\mathfrak{f}^{ik}$ on the one side and the tensor (B, E) or ϕ_{ik} on the other side. (It is what in elementary theory is called the material equations.) For, the only relationship one could think of, to wit $\mathfrak{f}^{ik} = \frac{1}{2}\epsilon^{iklm}\phi_{lm}$, makes the equations (A′), at least in the absence of current and charge ($\mathfrak{s}^k = 0$), a *consequence* of (B′) by identifying H with E and D with $-B$; which is entirely wrong *and could not be avoided by a different nomenclature.*

An alternative manner of getting at the required relationship would suggest itself in the light of later general developments.

We can easily explain it here directly. The quantity $\mathfrak{J}_2$, eqn. (2.11) is a scalar density. Hence the integral

$$I = \int \mathfrak{J}_2 dx^4,$$

taken over an invariantly fixed region, is an invariant. Now contemplate together with the original field ϕ_{ik} an 'infinitesimally neighbouring' field $\phi_{ik} + \delta\phi_{ik}$. The $\delta\phi_{ik}$, being, each of them, the difference of two tensors referring to the same point, are also a tensor field of the same character. Moreover

$$\delta I = \int \frac{\partial \mathfrak{J}_2}{\partial \phi_{ik}} \delta\phi_{ik} dx^4$$

is also an invariant, since it is the difference of the invariant I formed of $\phi_{ik} + \delta\phi_{ik}$ and that formed of ϕ_{ik}. From this it is easy to infer that the integrand is itself a scalar density; and since this holds for an arbitrary tensor $\delta\phi_{ik}$, we have that

$$\frac{\partial \mathfrak{J}_2}{\partial \phi_{ik}} = \text{contravariant skew tensor density of second rank.}$$

But a glance at (2.11) shows that it is the same as we got before by 'raising the subscripts with the help of the ϵ-density'. So this procedure is also of no avail.

PART II

AFFINELY CONNECTED MANIFOLD

CHAPTER III

INVARIANT DERIVATIVES

In order to find out (or, perhaps better, to agree upon) some natural way, by which to decide in an invariant manner whether and how a tensor varies from one point to the next, let us return to (2.16).

$$\frac{\partial A'_k}{\partial x'_i} = \frac{\partial x_l}{\partial x'_k}\frac{\partial x_m}{\partial x'_i}\frac{\partial A_l}{\partial x_m} + \frac{\partial^2 x_l}{\partial x'_i \partial x'_k} A_l. \qquad (3.1)$$

Suppose we had some reason to stipulate that A_k is to be regarded as 'really' constant, if all its 16 derivatives vanish in the original, the unprimed, frame (we thereby distinguish this frame *provisionally*). We will examine carefully what this statement amounts to in any other frame. In any other (the primed) frame it is, according to (3.1) expressed by

$$\frac{\partial A'_k}{\partial x'_i} - \frac{\partial^2 x_l}{\partial x'_i \partial x'_k} A_l = 0.$$

But in order to express it consistently in the primed frame, we had better replace the A_l by the A'_l, according to (2.15) (used the other way round). Thus

$$\frac{\partial A'_k}{\partial x'_i} - \frac{\partial x'_n}{\partial x_l}\frac{\partial^2 x_l}{\partial x'_i \partial x'_k} A'_n = 0.$$

Let us put, for abbreviation,

$$\frac{\partial x'_n}{\partial x_l}\frac{\partial^2 x_l}{\partial x'_i \partial x'_k} = \Gamma'^n{}_{ki}. \qquad (3.2)$$

Then the equations $$\frac{\partial A'_k}{\partial x'_i} - \Gamma'^n{}_{ki} A'_n = 0 \qquad (3.3)$$

express in an *arbitrary* frame the fact that the array of derivatives vanishes in the original, the unprimed, frame. Since the arbitrary transformation leading from the original to the primed frame may be specialized to be the identity, we have to say that the unprimed $\Gamma^n{}_{ik}$ are all zero. And, by the way, the same holds obviously for all frames which result from the original frame by a purely *linear* transformation of the x_k, since then the second derivatives in (3.2) all vanish.

And this is the only snag that remains and militates against the idea of general invariance: that one frame, or rather a set of frames, is distinguished by the assumption that in it, or in them, the Γ's all vanish. But this snag is very simply overcome: we just drop that assumption. This is a very important step which immediately leads to the concept of *affine connexion*.

So we now and hereafter do *not* define the Γ's by stipulating that they vanish in one particular frame and are given by (3.2) in any other one. We regard them as something of the general kind of a tensor field or a tensor-density field, but actually different from either—an array of functions which

(*a*) may be allotted arbitrary values in one particular frame, and

(*b*) are subject to a law of transformation that makes the following expression a tensor:

$$\frac{\partial A_k}{\partial x_i} - A_n \Gamma^n{}_{ki} = A_{k;i}. \tag{3.4}$$

The sign $A_{k;i}$ is a new notation introduced as an abbreviation for the expression on the left. We call the array of Γ's an *affine connexion* or, shorter, an affinity which we have by (*a*) imposed on our continuum. The $A_{k;i}$ is called the *invariant* derivative of A_k (with respect to the affinity $\Gamma^n{}_{ki}$), in contradistinction to the ordinary derivative $A_{k,i}$. Our previous consideration is to be regarded as a *special case*, viz. when *sub* (*a*) we choose the values zero for all the Γ's. From it we can easily infer that the requirement *sub* (*b*) will be satisfied, if we adopt for the $\Gamma^n{}_{ik}$ the following law of transformation: otherwise like a tensor corresponding to its three indices, but with an extra additional term, the expression on the left of (3.2). Thus

$$\Gamma'^n{}_{ik} = \frac{\partial x'_n}{\partial x_l} \frac{\partial x_r}{\partial x'_i} \frac{\partial x_s}{\partial x'_k} \Gamma^l{}_{rs} + \frac{\partial x'_n}{\partial x_l} \frac{\partial^2 x_l}{\partial x'_i \, \partial x'_k}. \tag{3.5}$$

The additional term is independent of the Γ's. It is thus the same for any affinity; it depends only on the relation between the two frames. It is responsible for the fact that the Γ's do not vanish in every frame even if they do so in one. An affinity is not a tensor. Its transformation formulae are *linear*, but *not homogeneous*.

The additional term is symmetrical with respect to the covariant indices *k* and *i* of Γ, and so is the whole transformation formula.

Symmetry with respect to the subscripts is therefore an invariant property of an affinity. (Antisymmetry is *not*!) If an affinity is non-symmetric, then in the transformation formula of its skew part, $\frac{1}{2}(\Gamma^n_{ik} - \Gamma^n_{ki})$, the non-homogeneous part drops out; this skew part therefore *is* a tensor. More generally, the fact that the non-homogeneous term is the same for *any* affinity has the following relevant consequences.

If we envisage *two* affine connexions Γ^k_{lm} and $\hat{\Gamma}^k_{lm}$ in the same continuum (as we may and very often do), then their *difference* $\Gamma^k_{lm} - \hat{\Gamma}^k_{lm}$ is always a tensor. In particular, if we have occasion to envisage an infinitesimal variation $\Gamma^k_{lm} + \delta\Gamma^k_{lm}$ of a given affinity Γ^k_{lm} (as we sometimes do), then the $\delta\Gamma^k_{lm}$ are a tensor. Inversely, of course, the sum of an affinity and a tensor T^k_{lm} is always an affinity.

The *sum* of two affinities is *not* an affinity, because in its transformation formula the critical term would have the factor 2. However, a linear aggregate of two affinities

$$\lambda\Gamma^k_{lm} + \mu\hat{\Gamma}^k_{lm}$$

is an affinity, if λ and μ are either fixed constants or invariants and

$$\lambda + \mu = 1.$$

Hence a non-symmetric affinity is always the sum of a symmetric affinity and a skew symmetric tensor of third rank, thus

$$\Gamma^k_{lm} = \tfrac{1}{2}(\Gamma^k_{lm} + \Gamma^k_{ml}) + \tfrac{1}{2}(\Gamma^k_{lm} - \Gamma^k_{ml}). \qquad (3.6)$$

The notion of a skew affinity is futile, because this property would not be independent of the frame.

Affinities are a second, or, if you like, a third, kind of relevant entities besides tensors and tensor densities. The notion of invariant derivative which we have introduced in (3.4) is not an absolute concept but refers to a certain affinity, which must be indicated. If more than one has been introduced and abbreviations (like the semicolon-notation used in (3.4)) are desirable, one must distinguish them by using various signs instead of the semicolon, as a colon, a vertical bar, etc., for the derivatives taken with respect to the several affinities.

We now want to extend the notion of invariant derivative to other tensors, first to the *contra*variant vector. A generalization is never compulsory, it is suggested by some simple guiding principle. In the present case it seems natural to demand

(1) that the ordinary rule of differentiating a product

$$\frac{\partial}{\partial x}(fg) = \frac{\partial f}{\partial x}g + f\frac{\partial g}{\partial x}$$

should apply also to the invariant differentiation of products of tensors;

(2) that in the case of an invariant the invariant derivative should be the ordinary derivative (since, after all, the gradient *is* a tensor—without supplementation!)

$$\phi_{;i} = \phi_{,i}.$$

To begin with a rather trivial remark, which however must be stated once and for all: since

$$A_k = \delta^l{}_k A_l,$$

the product rule alone tells us, that

$$A_{k;m} = \delta^l{}_k A_{l;m} + \delta^l{}_{k;m} A_l = A_{k;m} + \delta^l{}_{k;m} A_l.$$

And since this must hold for any vector, we must have

$$\delta^l{}_{k;m} = 0.$$

So the mixed unity tensor, regarded as a field, has the invariant derivative zero with respect to any affinity.

Now envisage the invariant product

$$A_k B^k$$

of two arbitrary vector-fields. According to the two guiding principles laid down we want

$$(A_k B^k)_{,i} = (A_k B^k)_{;i},$$

thus

$$A_k B^k{}_{,i} + A_{k,i} B^k = A_k B^k{}_{;i} + A_{k;i} B^k = A_k B^k{}_{;i} + (A_{k,i} - A_n \Gamma^n{}_{ki}) B^k.$$

By cancelling the terms $A_{k,i}B^k$, we get

$$A_k B^k{}_{;i} = A_k B^k{}_{,i} + A_n B^k \Gamma^n{}_{ki}.$$

This we write exchanging the dummies k, n in the last term

$$A_k(B^k{}_{;i} - B^k{}_{,i} - B^n \Gamma^k{}_{ni}) = 0.$$

Since A_k is an arbitrary vector:

$$B^k{}_{;i} = B^k{}_{,i} + B^n \Gamma^k{}_{ni}, \qquad (3.7)$$

This is the expression for the invariant derivative of a contravariant vector, the counterpart of (3.4) only in a slightly different notation, the comma indicating the ordinary derivative $\partial B^k/\partial x_i$.

If you have any doubt whether this $B^k{}_{;i}$ is a tensor, go back to an earlier equation from which (3.7) is derived, viz.

$$(A_k B^k)_{,i} = A_k B^k{}_{;i} + A_{k;i} B^k.$$

Here A_k is arbitrary and all terms save the first on the right are known to be vectors, hence $B^k{}_{;i}$ is a tensor.

One further remark: envisage

$$B^k{}_{,i} + B^n \Gamma^k{}_{in}. \tag{3.7a}$$

('We have made a mistake with the subscripts!') What is that? If Γ is symmetric, it is irrelevant. But what if it is not?

Well, this is a tensor all right, and it is an invariant derivative of B^k all right, only not just the one with respect to the affine connexion that we had envisaged, but with respect to another one, that results from it by exchanging the subscripts.

This is trivial. But it is worth while to observe also that no *logical* inconsistency would be involved, if we chose to adopt (3.7a) rather than (3.7) as defining the invariant derivative of a contravariant vector with respect to the *same* affinity, for which (3.4) is adopted in the covariant case. But, of course, with this choice the product rule would not hold for the semicolon-differentiation! However, this is only a side-remark, to which we give no consequence. That is we do adopt (3.7).

In the case of a general tensor

$$T^{kl\ldots}{}_{mn\ldots}$$

we apply similar considerations to the invariant

$$T^{kl\ldots}{}_{pq\ldots} A_k B_l \ldots F^p G^q \ldots,$$

with $A_k \ldots G^q \ldots$ arbitrary vectors; we thus get a result for the invariant derivative of T which we shall first describe in words, then write out. To the ordinary derivative there are additional supplementary terms, one for each index of T. Each such term consists of a (contracted) product of a component of T and a component of Γ, which product is formed exactly after the pattern of (3.4) or (3.7) respectively, whereby T is treated as though it had only that one index, all the others being disregarded, i.e. they are left unchanged in forming this particular product. Thus:

$$T^{kl\ldots}{}_{pq\ldots;i} = T^{kl\ldots}{}_{pq\ldots,i} + T^{nl\ldots}{}_{pq\ldots}\Gamma^{k}{}_{ni} + T^{kn\ldots}{}_{pq\ldots}\Gamma^{l}{}_{ni}$$
$$+\ldots - T^{kl\ldots}{}_{nq\ldots}\Gamma^{n}{}_{pi} - T^{kl\ldots}{}_{pn\ldots}\Gamma^{n}{}_{qi} - \ldots. \quad (3.8)$$

Notice that the differentiation index is always the *second* covariant index of Γ, the remaining two places being used for allocating the dummy and the one that is missing in T, where it has been replaced by the dummy. If this and the sign be remembered, the formula is easily memorized in the teeth of the bewildering dance of indices!

In order to extend invariant differentiation to densities, we supplement our guiding principle in the most natural way, viz.

(1) The product rule shall apply also if *one* factor is a density.

(2) The numerically invariant density ϵ^{iklm}, regarded as a field, shall have the derivative zero.

Let, for a scalar density $\mathfrak{S}$,

$$\mathfrak{S}_{;i} = \mathfrak{S}_{,i} + X,$$

where X is to be determined (we do not yet know what $\mathfrak{S}_{;i}$ means; we are about to define it!)

Now envisage any density $\mathfrak{T}^{\cdots}{}_{\ldots}$. If you divide it by an *arbitrary* scalar density $\mathfrak{S}$ you get a tensor, so you may put

$$\mathfrak{T}^{\cdots}{}_{\ldots} = \mathfrak{S}\,.\,T^{\cdots}{}_{\ldots},$$

where T (apart from being a tensor, not a density) is of the same character as $\mathfrak{T}^{\cdots}{}_{\ldots}$. Now we postulate

$$\mathfrak{T}^{\cdots}{}_{\ldots;i} = \mathfrak{S}\,.\,T^{\cdots}{}_{\ldots;i} + \mathfrak{S}_{;i}T^{\cdots}{}_{\ldots} = \mathfrak{S}\,.\,T^{\cdots}{}_{\ldots;i} + \mathfrak{S}_{,i}T^{\cdots}{}_{\ldots} + XT^{\cdots}{}_{\ldots}.$$

A brief consideration shows that the first and the second terms on the right together constitute just the 'ordinary terms', formed as if $\mathfrak{T}^{\cdots}{}_{\ldots}$ were a tensor. So we may write

$$\mathfrak{T}^{\cdots}{}_{\ldots;i} = \text{ordinary terms} + (X/\mathfrak{S})\,\mathfrak{T}^{\cdots}{}_{\ldots}.$$

Now, since $\mathfrak{S}$ was quite arbitrary and X depends on it alone, the factor $X/\mathfrak{S}$ must be independent of $\mathfrak{T}^{\cdots}{}_{\ldots}$ and can be determined from any special case. We determine it from the demand

$$0 = \epsilon^{klmn}{}_{;i} = 0 + \epsilon^{rlmn}\Gamma^{k}{}_{ri} + \epsilon^{krmn}\Gamma^{l}{}_{ri}$$
$$+ \epsilon^{klrn}\Gamma^{m}{}_{ri} + \epsilon^{klmr}\Gamma^{n}{}_{ri} + (X/\mathfrak{S})\,\epsilon^{klmn}.$$

In the summations over r only one term survives in each case, viz. the terms $r = k, l, m, n$ respectively. So we draw

$$0 = \epsilon^{klmn}(\Gamma^r_{ri} + X/\mathfrak{S}),$$

and thus

$$X/\mathfrak{S} = -\Gamma^r_{ri}.$$

Hence finally the extra additional term in the invariant derivative of *any density* $\mathfrak{T}^{\cdots}_{\cdots}$ reads

$$-\Gamma^r_{ri}\mathfrak{T}^{\cdots}_{\cdots}, \tag{3.9}$$

the indices on $\mathfrak{T}$ being *all* unchanged.

It is easy to show that not only do the guiding postulates we used lead to this unique determination of the invariant derivatives of tensors and densities, but inversely, with these definitions accepted, all those demands are actually fulfilled.

CHAPTER IV

SOME RELATIONS BETWEEN ORDINARY AND INVARIANT DERIVATIVES

Before we had introduced the notion of an affine connexion, we had learnt at the end of Chapter III that certain linear combinations of ordinary derivatives are tensors anyhow. They cannot, of course, lose this property by our imposing a connexion and introducing the notion of invariant derivatives with respect to it. However, the corresponding linear combinations of the invariant derivatives are tensors too—*a fortiori*, since the invariant derivatives are tensors even severally. We ask whether they are the same tensors or not.

We first study the cases labelled 1–4 in Chapter II.

In the case of the gradient of an invariant, there is no question: it was one of our guiding principles that for an invariant ϕ

$$\phi_{,i} = \phi_{;i}.$$

What about the curl of a covariant vector? From (3.4)

$$A_{k;i} - A_{i;k} = A_{k,i} - A_{i,k} - A_n(\Gamma^n{}_{ki} - \Gamma^n{}_{ik}). \qquad (4.1)$$

Thus the 'covariant curl' is the same as the ordinary curl, if and only if the affinity is symmetric.

To deduce and to remember general statements (unless you have to pass an exam, when you are frequently expected to memorize all sorts of stuff that nobody else knows by heart) is only useful if they have a frequent application. If the case does not arise very often, it is 'cheaper' to investigate it only as it arises. Non-symmetric affinities are rarely used. Hence, in order not to encumber the reader with gratuitous dead-weight, we restrict the further investigation *in this section* to symmetric affinities, with a strong emphasis, however, that our statements are definitely restricted *to them.*

It is not difficult to satisfy oneself by direct computation that the two cyclical divergences contemplated under (3) and (4), too, are the same, whether formed of the ordinary or of the invariant derivatives. That is to say (just as with the curl), if these formations turn up with 'semicolons', the simpler 'commas' may be substituted instead. The cases (4′)–(1′) referring to densitites do not require further investigation, for they are virtually the same as (4)–(1) on account of the general connexion between antisymmetric tensors and tensor densities of complementary rank. Indeed, after having ascertained, for example, that

$$A_{ik;l}+A_{kl;i}+A_{li;k} = A_{ik,l}+A_{kl,i}+A_{li,k}, \tag{4.2}$$

we need only put $\quad \mathfrak{A}^{lm} = \tfrac{1}{2}\epsilon^{lmik}A_{ik}.$

Then $$\frac{\partial \mathfrak{A}^{ml}}{\partial x_l} = \tfrac{1}{2}\epsilon^{mlik}\frac{\partial A_{ik}}{\partial x_l} = \pm(A_{ik,l}+A_{kl,i}+A_{li,k})$$

and $$\mathfrak{A}^{ml}{}_{;l} = \tfrac{1}{2}\epsilon^{mlik}A_{ik;l} = \pm(A_{ik;l}+A_{kl;i}+A_{li;k}),$$

so that $$\mathfrak{A}^{ml}{}_{;l} = \mathfrak{A}^{ml}{}_{,l}.$$

(The sign is in both cases that of the permutation *mlik*: we have used the fact that $\epsilon^{mlik}{}_{;s} = 0$, which, it will be remembered, was one of our 'guiding principles'.)

So we may take it that we have ascertained by direct inspection the equivalence of 'semicolon' and 'comma' in all eight cases. But these computations could be spared by an alternative proof which is shorter, is even more illuminating and applies to any of the eight cases severally. We illustrate it by the example of equation (4.2). To prove this equation directly, observe that both its first and its second member are known to be tensors. Moreover, from the general feature of the invariant derivative, their difference is a linear aggregate of components of $\Gamma^k{}_{lm}$. As a difference of two tensors it must be a tensor. *But there is no non-vanishing tensor, formed linearly from a symmetric connexion alone.* (This may seem a sweeping statement, of which the simple reason will be given in the last paragraph of this chapter.)

With a non-symmetric connexion there is; to wit its skew part $\tfrac{1}{2}(\Gamma^k{}_{lm}-\Gamma^k{}_{ml})$. That is the reason why our statements do not hold in this case.

The equality that arises from (4′)

$$\mathfrak{A}^k{}_{;k} = \mathfrak{A}^k{}_{,k} \tag{4.3}$$

entails a valuable rule for 'partial integration with respect to the invariant derivative' in four-dimensional, i.e. in space-time integrals. The rule is very simple. For the purpose of partial integration the semicolon can be treated as though it meant an ordinary derivative, provided one keeps sternly to the injunction that only an invariant density may figure as an integrand. This is proved as follows.

Suppose you had an integral of the following type

$$I = \int (A^{\cdots}{}_{\cdots})(B^{\cdots}{}_{\cdots})_{;k}\, dx^4, \tag{4.4}$$

where A and B are tensorial entities (tensors or densities) whose indices we have only indicated by dots. Now it is easy to see that since the integrand is to be a scalar density, the thing that arises from suppressing the covariant index k (i.e. the invariant differentiation) must be a contravariant vector density,† say $\mathfrak{A}^k$:

$$A^{\cdots}{}_{\cdots} B^{\cdots}{}_{\cdots} = \mathfrak{A}^k.$$

Moreover, from the rule for differentiating a product (one of our 'guiding principles'!)

$$(A^{\cdots}{}_{\cdots} B^{\cdots}{}_{\cdots})_{;k} = (A^{\cdots}{}_{\cdots})(B^{\cdots}{}_{\cdots})_{;k} + (A^{\cdots}{}_{\cdots})_{;k}(B^{\cdots}{}_{\cdots}).$$

Therefore $$I = \int [\mathfrak{A}^k{}_{;k} - (A^{\cdots}{}_{\cdots})_{;k}(B^{\cdots}{}_{\cdots})]\, dx^4.$$

On account of (4.3), the first part can be reduced to an integral over the (three-dimensional) 'surface'. This establishes our theorem.

In many applications (particularly to variational calculus) the first part vanishes and we have

$$I = -\int (A^{\cdots}{}_{\cdots})_{;k}(B^{\cdots}{}_{\cdots})\, dx^4.$$

This rule is not trivial. In A or B alone the (;) need not at all be equivalent to a simple (,). For instance (4.4) might read

$$I = -\int (A^{lm}{}_n)_{;k}(\mathfrak{B}^{nk}{}_{lm})\, dx^4,$$

where neither the tensor A nor the density $\mathfrak{B}$ need have any index symmetry.

If, as has sometimes been done with the idea of *simplifying* matters, the injunction of admitting only an invariant integrand is waived, the rule does not apply—with the sad effect of *complicating* the calculus enormously.

† Proof: $AB_{;k}$ is a scalar density. But ABF_k (with F_k an *arbitrary* vector) transforms exactly alike and is therefore also a scalar density. Hence AB is a contravariant vector density.

It is convenient to join here a further instance in which the invariant derivatives can—though in an entirely different order of ideas—be replaced by ordinary derivatives. One can always introduce a frame of coordinates so that the two kinds of derivatives coincide at any one particular point of the continuum. One can choose a frame for which all the Γ^i_{kl} vanish *at that point*. This can be shown as follows.

We mentioned in Chapter III that these components transform like those of a tensor, but with the first member of equation (3.2) added on to the familiar formula. Thus

$$\Gamma'^i_{kl} = \frac{\partial x'_i}{\partial x_r}\frac{\partial x_s}{\partial x'_k}\frac{\partial x_t}{\partial x'_l}\Gamma^r_{st} + \frac{\partial x'_i}{\partial x_m}\frac{\partial^2 x_m}{\partial x'_k \partial x'_l}. \tag{4.5}$$

We wish, by a suitable choice of the transformation, to make all the Γ' vanish at one point—say for simplicity at the point $x_k = 0$. We choose the transformation so that at this point the inverse transformation has the analytical development

$$x_k = x'_k + \tfrac{1}{2}a^k_{lm}x'_l x'_m + \ldots,$$

where we assume $a^k_{ml} = a^k_{lm}$, because obviously nothing would be gained if we did not. We get from (4.5), at the point $x_k = x'_k = 0$,

$$\Gamma'^i_{kl} = \Gamma^i_{kl} + a^i_{kl} = 0,$$

provided that we choose

$$a^i_{kl} = -(\Gamma^i_{kl})_{\text{at } x_k=0}.$$

Such a choice of coordinates is called a geodesic frame (or geodesic coordinates). The full verbal description is, of course, a frame geodesic at a certain point for a certain symmetric affinity.

Clearly, if the Γ's are not symmetric, they cannot be 'transformed away', not even at one point. That is small wonder, because the skew part is a tensor, which you cannot make vanish in any frame, unless it vanishes in every frame. On the other hand, it is not required for this purpose that Γ be symmetric at large, that it be 'a symmetric affinity'. It must only be symmetric at the point in question. (This, if it happens, is obviously an invariant feature. It means that the skew tensor happens to vanish at the point in question.)

A geodesic system is often very convenient for computations, but one must always keep in mind that it simplifies matters only at one point, nowhere else, not even in the neighbouring points. I mean to say, the derivatives of the Γ's do not vanish, and one must be careful in the course of a computation not to drop a Γ which would afterwards come in for being differentiated with respect to a coordinate.

A further consequence is the truth of our 'sweeping statement' on p. 35, that there is no non-vanishing tensor with components that are linear aggregates of the components of a *symmetric* affinity. Indeed such linear aggregates, no matter what the coefficients are, must all vanish in a geodesic frame, and therefore, if they form a tensor, in every frame. Since this consideration applies to every point of the continuum, a tensor of the said description would vanish identically.

CHAPTER V

THE NOTION OF PARALLEL TRANSFER

There is an alternative way of introducing the notion of affine connexion and invariant derivative. In view of the fundamental character of these notions in *all* our considerations we shall indicate this alternative.

$^{\circ}Q_{(x_i+dx_i)}$

$\overset{\circ}{P}_{(x_i)}$

The array of derivatives $\partial A^k/\partial x_i$ does not constitute an invariant entity, because they are formed by the 'inadmissible' procedure of subtracting the vector A^k at P from the vector $A^k + dA^k$ at *another* point, namely at the neighbouring point Q with coordinates $x_i + dx_i$ (x_i being those of P). Their difference dA^k *is not a vector.* Hence according to the correct formula

$$dA^k = \frac{\partial A^k}{\partial x_i} dx_i,$$

$\partial A^k/\partial x_i$ *cannot* be a tensor, since dx_i *is* a vector.

To remedy this defect you must, for the purpose of forming a derivative, subtract from $A^k + dA^k$ not the vector A^k at P, but some vector at Q, which for this purpose so to speak takes the place of A^k, i.e. it plays the part of the 'unchanged' or 'original' value of the function in ordinary differentiation. In other words, you must stipulate by definition what change in the components of A^k, on proceeding from P to Q, you regard as 'no change'. (The simple suggestion that just *no* change in the numerical values of the components should represent 'no change' of the geometrical entity is not good enough, because this stipulation is not independent of the frame.)

Let this 'substitute at Q for A^k at P' or this 'by definition unchanged entity at Q, corresponding to the entity A^k at P', be called $A^k + \delta A^k$. Naturally δA^k, being the difference of a vector at Q (viz. $A^k + \delta A^k$) and a vector at P (viz. A^k), is also not a vector, just as and for the same reasons as dA^k is not; but $dA^k - \delta A^k$ is one.

δA^k must be made to depend on the two vectors A^k and dx_i and cannot depend on anything else. Moreover, it must be made to vanish with either of these two vectors. A homogeneous *linear* dependence on both A^k and dx_i suggests itself as being the simplest ruling. Hence we contemplate a bilinear form of these two vectors

$$\delta A^k = -\Gamma^k{}_{li} A^l \, dx_i, \tag{5.1}$$

the Γ's being an array of 64 coefficients, functions of the coordinates, newly introduced from the point of view taken in this chapter (but, of course, we have conformed the notation to that of Chapter III).

Since A^k and dx_i are vectors, but δA^k is not, the Γ's do not constitute a tensor. They follow the linear but non-homogeneous transformation law that was already indicated in Chapter III, equation (3.5). That can easily be shown, by demanding that the association of the vectors A^k in P and $A^k + \delta A^k$ in Q should subsist on coordinate transformation. We skip the proof here.

The vector

$$A^k + \delta A^k \equiv A^k - \Gamma^k{}_{li} A^l \, dx_i \tag{5.2}$$

is called the parallel-displaced (or parallel-transferred) vector. The invariant derivative $A^k{}_{;i}$ is now defined thus:

$$\begin{aligned} A^k{}_{;i} &= \frac{(A^k + dA^k - A^k - \delta A^k)_{\text{only } dx_i \neq 0}}{dx_i} \\ &= \frac{(dA^k - \delta A^k)_{\text{only } dx_i \neq 0}}{dx_i} \\ &= A^k{}_{,i} + \Gamma^k{}_{li} A^l \end{aligned} \tag{5.3}$$

which conforms exactly with (3.7).

The Γ's establish a linear one-to-one *mapping* of the 'vector-hedgehogs' in neighbouring points on each other. A linear mapping between two sets of four functions requires 16 arbitrary coefficients, but since there are ∞^4 neighbouring points, we have 64 of them. In elementary geometry a linear transformation of the coordinates is called an affine transformation. Geometrical figures that are turned into each other by such a transformation are called affinely related—e.g. a sphere and a concentric ellipsoid in three dimensions. The geometry 'under this group' is called 'affine geometry': it contemplates only such properties as are invariant under affine transformation and are thus the same for any two affinely related geometrical figures, e.g. all ellipsoids, including the sphere, are considered as the same figure.

That is how the name affine connexion or affinity has arisen. It is *not* a good name. For two reasons. First, already in a non-connected manifold the general coordinate transformation entails an arbitrary affine transformation at any point. On the other hand, the Γ's indicate just *not* an arbitrary affine relationship between neighbouring points, but they distinguish once and for all a particular one.

So much for explaining and criticizing the terminology.

Along the ideas of the present chapter we should now proceed to define analogously to (5.1) the parallel-displacement of any tensor or tensor density, using as guiding principles (1) that a product is displaced by displacing all its factors, (2) that an invariant is not to change on displacement, (3) that the ϵ-density is not to change on displacement. In this way one easily deduces that the change $\delta T^{kl\ldots}{}_{pq\ldots}$ of any *tensor* on displacement must be given by the additional terms of (3.8), multiplied by $(-dx_i)$. In the case of a tensor *density* of any type the term (3.9), multiplied by $(-dx_i)$, has to be added. The invariant derivatives are then defined in exact analogy with (5.3) and are, of course, those given in (3.8), with the 'amendment' (3.9) in the case of a tensor density.

Along the line of thought of the present chapter there is nothing to suggest that $\Gamma^k{}_{lm}$ should be symmetric in l and m. And since also in the previous chapters there was nothing to *enforce* this assumption, we shall not make it in general, but regard the symmetric affinity as a special (though very frequent and important) case, to be indicated every time we deal with *it*, and not with the general case.

The aspect of an affinity as constituting a system of parallel-transfer is the more fundamental one. Thus the approach which puts it in the first place and the notion of invariant derivation in the second place is more fundamental than that followed in Chapter III.

Simple and obvious as it is, we ought yet to stress the fact that the parallel-transfer of a zero tensor or zero tensor density is always again zero. This has the consequence that any tensor equation is always preserved on parallel-transfer, just as it is on coordinate transformation. Indeed it can always be reduced to the statement that a certain linear aggregate of products of tensors is equal to the zero tensor of the same character (i.e. the same as regards co- and contravariant indices and whether it is a density or a tensor).

But please take care! The equation holds at the other point for the parallel-transferred tensors. But the tensors *may* be field-tensors. And their actual values at the other point need not and, as a rule, will not be those obtained by parallel-transfer. So the equation need not hold for the field-tensors at the neighbouring point!

That is the more necessary to underline, because in point of fact they very often *do* hold there—and everywhere; namely when they are field-equations which have been explicitly established or declared to hold at every point.

CHAPTER VI

THE CURVATURE TENSOR

THE QUESTION OF INTEGRABILITY

Easily the most interesting and vital point that occurs in the study of affine connexion and parallel-transfer is this. If you envisage a tensorial entity, e.g. a contravariant vector A^k at a point P (not necessarily a member of a 'field'), and carry it by *continual* parallel-transfer around a closed circuit C back to P, the entity does not in general (i.e. for an arbitrary affinity) return to its original value, but you arrive at P with a different entity, say with a vector $\hat{A}^k \neq A^k$.

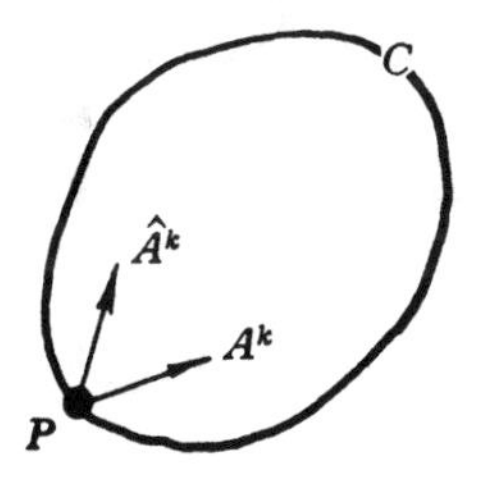

It is only a different way of expressing the same fact to say that the result of transferring A^k from P to any other point Q will, as a rule, depend on the path. For example, if the transfer of A^k over S to Q yields $A^k(S)$ (say), transfer via R will lead to something else, say $A^k(R)$ in Q, $\neq A^k(S)$. For obviously the transfer there and back along the *same* curve is reversible; so *if* $A^k(R) = A^k(S)$, then $\hat{A}^k = A^k$, and vice versa.

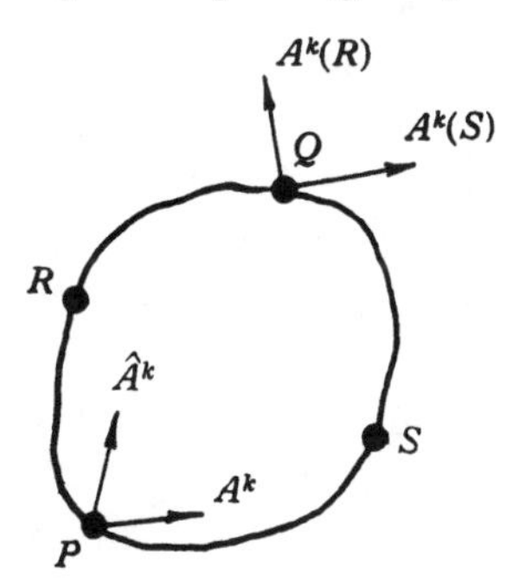

If the transfer is independent of the path for any vector A^k (and then, as we shall see, for any tensorial entity), we call the affine connexion integrable—or, possibly, integrable within a certain region, if path-independence holds only for points and paths within a certain region. Whether an affinity is integrable or not is decided, as we shall see presently, by the vanishing or otherwise of a certain tensor of the fourth rank, which is called the curvature tensor or the Riemann-Christoffel tensor and plays the central rôle in *all* theories of the structure of space-time, so that we shall be dealing with it incessantly in all the considerations to follow. And, of course, it is the non-integrable case when this tensor does not vanish (curved manifold) that presents the greater interest.

For the moment, however, we shall gather information on the integrable case by a more direct method. In this case let h^ν be a single vector (not a field) at a point P. Since, now, the parallel-transfer does not depend on the path, we can 'spread' h^ν into a field by parallel-transfer from P, defining the field vector h^ν by the condition that its actual change on proceeding from one point to the next shall be equal to the change on parallel-transfer. Using our former notation, this is expressed by

$$dh^\nu = \delta h^\nu$$

or written more elaborately

$$\frac{\partial h^\nu}{\partial x_\lambda} dx_\lambda = -\Gamma^\nu{}_{\alpha\lambda} h^\alpha dx_\lambda.$$

Since that is to hold for any dx_λ, it amounts to subjecting h^ν to the differential equations

$$\frac{\partial h^\nu}{\partial x_\lambda} = -\Gamma^\nu{}_{\alpha\lambda} h^\alpha, \tag{6.1}$$

together with the initial condition that at P the h^ν are to take the values given there. The slight inconsistency that we have first used h^ν to indicate the vector at P, and now use it for the field-vector obtained by integrating (6.1), does not matter. Indeed, after 'spreading', the point P is no longer distinguished; the field-vector at any point is obtained by parallel-transfer from any other point along any path connecting the two points. This state of affairs (and similar ones in the case of other tensor-fields) is suitably expressed by saying: the affinity Γ carries (viz. by parallel-transfer) the field-vector h over into itself. Notice by the way that according to (3.4) the equations (6.1) simply state that h has a vanishing invariant derivative.

Now let us do the same thing with four linearly independent vectors at P, which we *label* by a *subscript a*, thus $h^\nu{}_a$, with $a = 1, 2, 3, 4$. We might just as well use four different letters, as $h^\nu, g^\nu, f^\nu, j^\nu$ to indicate our four vectors, instead of that subscript, *which must not be confused with a tensor-index.* The numbering is more convenient, because it is convenient for the purpose of the present investigation to extend the summation convention to this subscript, when it appears twice (though it will always be written as a *sub*script, being a mere label).

We have chosen the four vectors $h^\nu{}_a$ linearly independent at P. This linear independence will obviously hold for the field-vectors at any point, simply because a linear relation

$$c_1 h^\nu_1 + c_2 h^\nu_2 + c_3 h^\nu_3 + c_4 h^\nu_4 = 0 \tag{6.2}$$

(the c_k being numerical constants, not all $= 0$) would be conserved by parallel-transfer and can therefore, not hold at any point of the field, since it shall not hold at P.

So we have now four vector-fields governed by the relations

$$\frac{\partial h^\nu{}_a}{\partial x_\lambda} = -\Gamma^\nu{}_{\alpha\lambda} h^\alpha{}_a. \tag{6.3}$$

These are 64 linear non-homogeneous equations with non-vanishing determinant,† from which the 64 quantities Γ can be determined everywhere. We do this in the usual way. The normalized minors in the determinant $h^\nu{}_a$ we call $h_{\nu a}$ (by normalized we mean divided by the determinant). Then

$$h_{\rho a} h^\nu{}_a = \delta^\nu{}_\rho. \tag{6.4}$$

(To justify the notation: $h_{\rho a}$ for a fixed a *is* a covariant vector-field. This follows from the consideration that

(i) the preceding equations are to hold in every frame of reference, the $h_{\rho a}$ always meaning the normalized minors;

(ii) the $h_{\rho a}$ are uniquely determined by these equations;

(iii) the equations *are* preserved, if the $h_{\nu a}$ are transformed as vector components.)

We still note the fact that, inversely, the $h^\nu{}_a$ are the normalized minors of the determinant $h_{\nu a}$. Now 'multiply' (i.e. multiply and sum over a) the equation (6.3) by $h_{\rho a}$, then you get by (6.4)

$$\Gamma^\nu{}_{\rho\lambda} = -h_{\rho a} \frac{\partial h^\nu{}_a}{\partial x_\lambda}. \tag{6.5}$$

This shows that ***integrability*** is a severe restriction on an affinity. It makes it possible to express the 64 functions Γ by the components of four vector-fields, i.e. by only 16 functions.

† It is not difficult to show that the determinant could vanish at a point if, and only if, a relation like (6.2) would hold at this point.

The preceding representation of an integrable affinity has a particularly simple consequence, if Γ is *symmetric* in its two covariant indices. So let us now assume also

$$\Gamma^{\nu}{}_{\rho\lambda} = \Gamma^{\nu}{}_{\lambda\rho}. \tag{6.6}$$

Now from (6.4) the equations (6.5) can be written equivalently

$$\Gamma^{\nu}{}_{\rho\lambda} = h^{\nu}{}_{a} \frac{\partial h_{\rho a}}{\partial x_{\lambda}}. \tag{6.7}$$

Hence, from the assumption (6.6)

$$h^{\nu}{}_{a}\left(\frac{\partial h_{\rho a}}{\partial x_{\lambda}} - \frac{\partial h_{\lambda a}}{\partial x_{\rho}}\right) = 0$$

and thus, since the determinant does not vanish,

$$\frac{\partial h_{\rho a}}{\partial x_{\lambda}} - \frac{\partial h_{\lambda a}}{\partial x_{\rho}} = 0, \tag{6.8}$$

in words: the four covariant vector-fields $h_{\rho a}$ are curl-free.

This enables us to introduce a new frame of coordinates in the following way. To a fixed point P we assign the coordinates 0, 0, 0, 0. To any other point Q we assign the coordinates

$$y_a = \int_P^Q h_{\rho a} dx_{\rho} \quad (a = 1, 2, 3, 4). \tag{6.9}$$

Indeed from general analysis it is known that (6.8) are the necessary and sufficient conditions for this line integral to be independent of the path. Moreover, the derivatives of the y's with respect to the x's are obviously

$$\frac{\partial y_a}{\partial x_{\rho}} = h_{\rho a}.$$

Forming the normalized minors on both sides, you get

$$\frac{\partial x_{\rho}}{\partial y_a} = h^{\rho}{}_{a}.$$

Hence (6.7) can be expressed thus:

$$\Gamma^{\nu}{}_{\rho\lambda} = \frac{\partial x_{\nu}}{\partial y_a} \frac{\partial^2 y_a}{\partial x_{\rho} \partial x_{\lambda}}. \tag{6.10}$$

Comparing this with (3.2) and considering the context there, we infer:

Our integrable symmetric Γ-affinity can be regarded as the result of transforming the affinity with vanishing components in the y-frame from the y-frame to the x-frame. Or, putting it the other way round: all the components of our affinity vanish when transformed to the y-frame.

Thus a *symmetric* integrable affinity is a very simple thing, it can always be 'transformed to zero'.

It is small wonder that this theorem is restricted to the symmetric case. For remember that the general non-symmetric affinity can be split up into the sum of a symmetric one and a skew-symmetric tensor of the third rank. These two entities keep cleanly separated on transformation. And, of course, the skew part, being a tensor, can never be annihilated by transformation, unless it vanishes at the outset. All we can say in general is that the injunction of being integrable entails in every case that the components of the affinity are expressible by the 16 components of four vector-fields. This is even a greater reduction in the case of a non-symmetric affinity (which has otherwise 64 independent components) than for a symmetrical one, which in general has 40.

THE CURVATURE TENSOR

Given the 64 functions $\Gamma^{\nu}_{\alpha\lambda}$ that constitute an affine connexion, it would be difficult to decide directly whether or not they can be expressed in the form (6.5). We are therefore out for a criterion of integrability that can be applied straight away to the Γ-field when it is given.

To derive a necessary condition is very easy. For integrability to obtain, the equations (6.1) must admit of a solution for the vector-field h^{ν} with arbitrary initial values. Then surely the mixed second derivatives of the field-components, formed in two different ways, must agree. Hence we must have

$$\begin{aligned} 0 &= -\frac{\partial}{\partial x_{\mu}}(\Gamma^{\nu}_{\alpha\lambda}h^{\alpha}) + \frac{\partial}{\partial x_{\lambda}}(\Gamma^{\nu}_{\alpha\mu}h^{\alpha}) \\ &= \left(-\frac{\partial \Gamma^{\nu}_{\alpha\lambda}}{\partial x_{\mu}} + \frac{\partial \Gamma^{\nu}_{\alpha\mu}}{\partial x_{\lambda}}\right)h^{\alpha} - \Gamma^{\nu}_{\alpha\lambda}\frac{\partial h^{\alpha}}{\partial x_{\mu}} + \Gamma^{\nu}_{\alpha\mu}\frac{\partial h^{\alpha}}{\partial x_{\lambda}}. \end{aligned} \qquad (6.11)$$

Using equation (6.1) to express the first derivatives, we get (mind the notation for the dummy indices!)

$$\begin{aligned} 0 &= \left(-\frac{\partial \Gamma^{\nu}{}_{\alpha\lambda}}{\partial x_{\mu}} + \frac{\partial \Gamma^{\nu}{}_{\alpha\mu}}{\partial x_{\lambda}} + \Gamma^{\nu}{}_{\beta\lambda}\,\Gamma^{\beta}{}_{\alpha\mu} - \Gamma^{\nu}{}_{\beta\mu}\,\Gamma^{\beta}{}_{\alpha\lambda} \right) h^{\alpha} \\ &= B^{\nu}{}_{\alpha\lambda\mu} h^{\alpha}, \end{aligned} \tag{6.12}$$

where we have abbreviated the bracket expression by the B symbol. Since this holds for arbitrary h^{α}, we must have

$$B^{\nu}{}_{\alpha\lambda\mu} = 0 \tag{6.13}$$

in every frame of reference (which justifies the presumption that the B's form a tensor). So this is a necessary condition for integrability.

To comprehend that it is also *sufficient*, notice that our demand (6.11) *would* amount exactly to the well-known necessary and sufficient condition for the Pfaff-differential

$$-\Gamma^{\nu}{}_{\alpha\lambda} h^{\alpha} dx_{\lambda} \tag{6.14}$$

(which is the δh^{ν} on parallel transfer along dx_{λ}) to be a *complete* differential, *if* the h^{α} were given functions of the coordinates. This they are not, and it would be a vicious circle to anticipate it, for that is just what we want to prove.

However, in the small vicinity of any point the equations (6.1) together with (6.11) do suffice to determine the first and second derivatives of the h^{α} uniquely and without contradiction. Hence, given the initial values at a point, we can determine the field h^{α} in a small neighbourhood of this point, including quantities of the second order with respect to the differences of coordinates. If you take in (6.14) these functions h^{α}, it becomes a Pfaff-differential whose integrability conditions (6.11) are fulfilled including the first order (only the first, because they contain the derivatives of the h^{α}). The integral of (6.14) taken around any small circuit contained in that neighbourhood will therefore vanish with an accuracy including the second order, that is to say it can at most reach the third order.

Now envisage two infinitely neighbouring curves leading from P to a distant point Q. From the initial values h^{α} given at P we first build up the field h^{α} in the vicinity of P, choose a neighbouring point on C well within this vicinity, do the same there and so on, until we reach Q. In all this we take care that the sum-total of our small regions should cover also the curve C'. Then, first, the h^α reached at Q can be 'wrong' (as compared with *exact* transfer) only by quantities of the first order. Secondly, if you dissect the strip between C and C' into small surface-elements, the circuit integral around such a one will not exceed the third order, from which you easily deduce that the line-integral along C and that along C' cannot differ by more than the second order.

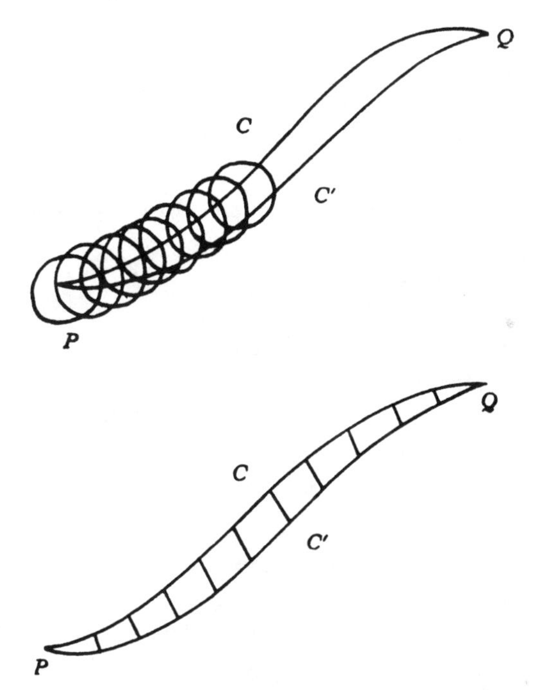

In the same way you can distort the curve C in small steps into any other curve joining P and Q. The line-integrals will then only differ by a quantity of the first order.

By making the subdivisions and the steps of distortion smaller and smaller, you arrive in the end at proving that (6.13) is also *sufficient* for the transfer of a contravariant vector to be integrable. To prove this was our aim.

The path-independence of the transfer for a covariant vector follows from that for the contravariant one because the invariant

$B_k A^k$ remains unchanged, for a fixed B_k and any A^k. In a similar way you show the same for higher tensors, and tensor densities.

If (6.13) is not fulfilled, any vector h^α can still be 'spread' according to (6.1) in the neighbourhood of a point, but only including quantities of the first order. Using this in (6.14), the integral of this differential around an infinitesimal circuit can easily be indicated from the mathematical theorem of Stokes. For the infinitesimal quadrangle with corners x_k, $x_k + dx_k$, $x_k + dx_k + \delta x_k$, $x_k + \delta x_k$ one obtains

$$B^{\nu}{}_{\alpha\lambda\mu} h^\alpha dx_\lambda \delta x_\mu \tag{6.15}$$

as the quantities by which the components h^ν change by parallel-transfer around that quadrangle. *But this theorem cannot be extended to a finite circuit*—simply because in this case there is in a finite region no field h^α to which it would apply!

Since (6.15), being the difference of two vectors at the point x_k, is itself a vector, and that for arbitrary vectors h^α, dx_α, δx_α, it follows that B is a tensor of fourth rank. An alternative proof of this important fact results from producing the B-components in a different way, that is in itself of interest, viz. by *commuting* two invariant differentiations in any tensor. For example, for a contravariant vector-field you get by straightforward computation:

$$A^{\nu}{}_{;\lambda;\mu} - A^{\nu}{}_{;\mu;\lambda} = -B^{\nu}{}_{\alpha\lambda\mu} A^\alpha - (\Gamma^{\beta}{}_{\lambda\mu} - \Gamma^{\beta}{}_{\mu\lambda}) A^{\nu}{}_{;\beta}. \tag{6.16}$$

The second term on the right vanishes when Γ is a symmetric affinity, for it depends only on its skew part. The latter, however, is a tensor in any case, and so, since the A-field is arbitrary, the tensor property of B follows immediately.

In the most general case (non-symmetric affinity) the B-tensor has only the one obvious symmetry of being skew in the last two subscripts, λ and μ in our present notation. It is easily seen that it then has $4 \times 4 \times 6 = 96$ independent components.

If Γ is symmetric, we gather from the explicit expression (6.12) that B has in addition the cyclical symmetry

$$B^{\nu}{}_{\alpha\lambda\mu} + B^{\nu}{}_{\lambda\mu\alpha} + B^{\nu}{}_{\mu\alpha\lambda} = 0. \tag{6.16a}$$

To count the independent components in this case, first give ν a fixed value. Then take for the subscripts a definite triplet $\alpha \neq \lambda \neq \mu$. Of the *three* independent components, which the plain skewness in the last couple leaves in this case, *one* can be expressed by the other two in virtue of the afore-standing relation. That leaves two for every triplet $\alpha \neq \lambda \neq \mu$, thus *eight* for the four different triplets. In the event of *two* out of the three α, λ, μ being equal (for which there are $2 \times 6 = 12$ possibilities), already the plain skewness leaves only one independent component in every case (for example $B^{\nu}{}_{212} = -B^{\nu}{}_{221}$ and $B^{\nu}{}_{122} = 0$). Moreover, the cyclical condition is in this case automatically fulfilled; it only restates the plain skewness. Hence we have *twelve* further components—and that is all. So we have $8+12 = 20$, for fixed ν, and thus 80 on the whole; the Riemann-Christoffel-tensor of a symmetric affinity has eighty independent components. We shall later come to know a case of still higher symmetry and special importance, when the number reduces to twenty only.

The B-tensor can be contracted with respect to (ν, α), or (ν, λ) or (ν, μ), but the latter two, on account of the skew symmetry, are not essentially different. We collect from (6.12) the expression of the B-tensor and add those of its two contractions, using a different (more customary) labelling from the one at which we had arrived by chance:

$$\left.\begin{aligned}
B^{i}{}_{klm} &= -\frac{\partial \Gamma^{i}{}_{kl}}{\partial x_m} + \frac{\partial \Gamma^{i}{}_{km}}{\partial x_l} + \Gamma^{i}{}_{\alpha l}\,\Gamma^{\alpha}{}_{km} - \Gamma^{i}{}_{\alpha m}\,\Gamma^{\alpha}{}_{kl},\\
R_{kl} = B^{\beta}{}_{kl\beta} &= -\frac{\partial \Gamma^{\alpha}{}_{kl}}{\partial x_\alpha} + \frac{\partial \Gamma^{\alpha}{}_{k\alpha}}{\partial x_l} + \Gamma^{\beta}{}_{\alpha l}\,\Gamma^{\alpha}{}_{k\beta} - \Gamma^{\beta}{}_{\alpha\beta}\,\Gamma^{\alpha}{}_{kl},\\
&\qquad\text{(Einstein-tensor)}\\
S_{lm} = B^{\beta}{}_{\beta lm} &= -\frac{\partial \Gamma^{\alpha}{}_{\alpha l}}{\partial x_m} + \frac{\partial \Gamma^{\alpha}{}_{\alpha m}}{\partial x_l}.\\
&\qquad\text{(Second contraction)}
\end{aligned}\right\} \qquad (6.17)$$

In the last formula it is remarkable that the simple 'curl' of the four components $\Gamma^{\alpha}{}_{\alpha l}$ (which are *not* a covariant vector) turns out to be a tensor.

R_{ki} is *not* in general symmetric, not even when Γ is (the second term is in the way even then). In any case, of course, its symmetric and its skew parts are tensors, too. Moreover, its skew part reduces to $-S_{kl}$ if Γ is symmetric. Hence in the case of a symmetric affine connexion there is virtually only one relevant contraction of the B-tensor, viz. the Einstein-tensor.

Having the rather complicated formulae for B and R at hand, we should like to add a very useful simple theorem concerning the changes these tensors undergo when the connexion is slightly varied: $\Gamma^k{}_{lm} \rightarrow \Gamma^k{}_{lm} + \delta\Gamma^k{}_{lm}$. It will be remembered that $\delta\Gamma^k{}_{lm}$ (in contradistinction to $\Gamma^k{}_{lm}$ itself) *is* a tensor. The following formula can be made good by straightforward computation

$$\delta B^i{}_{klm} = -(\delta\Gamma^i{}_{kl})_{;m} + (\delta\Gamma^i{}_{km})_{;l} + (\Gamma^\alpha{}_{ml} - \Gamma^\alpha{}_{lm})\,\delta\Gamma^i{}_{k\alpha} \quad (6.18)$$

and, by contraction

$$\delta R_{kl} = -(\delta\Gamma^\alpha{}_{kl})_{;\alpha} + (\delta\Gamma^\alpha{}_{k\alpha})_{;l} + (\Gamma^\alpha{}_{\beta l} - \Gamma^\alpha{}_{l\beta})\,\delta\Gamma^\beta{}_{k\alpha}. \quad (6.19)$$

These expressions are particularly convenient in the case of a symmetric connexion, where the last term in each of them vanishes.

CHAPTER VII

THE GEODESICS OF AN AFFINE CONNEXION

Given an affinity Γ^k_{lm}, let us envisage at a point $P(x_k)$ a line element dx_k, leading from P to the point $P'(x_k+dx_k)$. Transfer dx_k (being a vector in P) according to the connexion Γ from P to P', and let the result be $d'x_k$ (being a vector in P').

Transfer this vector from P' to $P''(x_k+dx_k+d'x_k)$. The result, $d''x_k$ transfer to $P'''(x_k+dx_k+d'x_k+d''x_k)$ and so on.

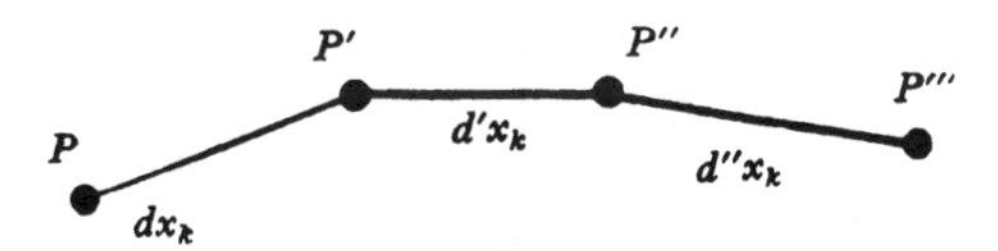

In this way you obtain a polygonal track that in the limit of 'true infinitesimals' and an infinitely increasing number of steps approaches a curve, which, by the way, could also be traced 'backwards' from P in the direction $-dx_k$, and which obviously has the following properties:

(i) If we transfer a finite contravariant vector indicating the direction of the curve at any of its points P, from P along the curve to any other point Q, we obtain a vector indicating the direction of the curve at Q. (By indicating the direction, we mean: being tangential to the curve or having components proportional to the increments dx_k along the curve.)

(ii) Our construction affords a natural standard for comparing the *lengths* of any two sections of this curve (*natural* with respect to the connexion Γ), namely the ratio of the 'number of steps' involved in each of them, or, to be accurate, the limiting value of this ratio.

One such curve issues from every one of the ∞^4 points in every (∞^3) direction, so there are ∞^6 such curves in all (since a curve contains ∞^1 points). Generally speaking, the ∞^3 curves issuing from a given point cover a certain *finite* neighbourhood of the point just once and there will be just one curve connecting two given points P and Q. These curves are called *geodesics*. We proceed to study them analytically.

Any curve can in many ways be represented by giving its four coordinates as functions of a continuous parameter λ. ('In many ways' means only that instead of λ we may choose any continuous monotonical function of λ.) If we do that, the vector $dx_k/d\lambda$ indicates at every point the direction of the curve in the way explained above.

To comply with (i) above, we demand that this vector, when parallel-transferred to $x_k + dx_k$, shall be proportional to the value of the vector that you encounter at the neighbouring point:

$$\frac{dx_k}{d\lambda} - \Gamma^k{}_{lm}\frac{dx_l}{d\lambda}dx_m = M\left(\frac{dx_k}{d\lambda} + \frac{d^2x_k}{d\lambda^2}d\lambda\right),$$

where M is some number. Dividing by $d\lambda$ we have

$$M\frac{d^2x_k}{d\lambda^2} + \Gamma^k{}_{lm}\frac{dx_l}{d\lambda}\frac{dx_m}{d\lambda} = \frac{1-M}{d\lambda}\frac{dx_k}{d\lambda}.$$

In order that this should make sense, M must differ from unity only by the order of $d\lambda$, which is pretty understandable. It can therefore be replaced by 1 on the left. On the right we must allow $1-M$ to depend on λ and so we write $\phi(\lambda)\,d\lambda$ for it. Thus

$$\frac{d^2x_k}{d\lambda^2} + \Gamma^k{}_{lm}\frac{dx_l}{d\lambda}\frac{dx_m}{d\lambda} = \phi(\lambda)\frac{dx_k}{d\lambda}. \tag{7.1}$$

This incorporates our first demand. But is λ the natural measure of the length along the curve, to which we pointed in (ii) above? Hardly, for its choice was to a large extent arbitrary. Let us see what (7.1) becomes, if we alter our choice, taking $s(\lambda)$ instead. We easily obtain

$$\frac{d^2x_k}{ds^2} + \Gamma^k{}_{lm}\frac{dx_l}{ds}\frac{dx_m}{ds} = \frac{\phi s' - s''}{s'^2}\frac{dx_k}{ds}, \tag{7.2}$$

where s', s'' mean the derivatives with respect to λ.

We can make the second member vanish altogether and give our equation the form

$$\frac{d^2x_k}{ds^2} + \Gamma^k{}_{lm}\frac{dx_l}{ds}\frac{dx_m}{ds} = 0, \tag{7.3}$$

if, and only if, we demand that $\phi s' - s'' = 0$, of which the general solution is

$$s = \int^\lambda \exp\left[\int^\lambda \phi(u)\,du\right]d\lambda. \tag{7.4}$$

Thus for the simplified form (7.3) of the differential equation of the geodesic to obtain, the choice of the variable s is determined up to a linear transformation with constant coefficients of the type $\hat{s} = as + b$, a liberty embodied in the lower limits of the two integrations in (7.4).

Equation (7.3) clearly enunciates that dx_k/ds is parallel-transferred along the geodesic. The vector $\left(\frac{dx_k}{ds}\right)_{\text{at } Q}$ is the parallel displaced of $\left(\frac{dx_k}{ds}\right)_{\text{at } P}$. The same holds for the infinitesimal vectors $(dx_k)_{\text{at } P}$ and $(dx_k)_{\text{at } Q}$, *if ds is given the same infinitesimal value at both points.* Hence ds is a measure of the length of an infinitesimal section and $\int ds$ a measure of a finite section of the geodesic in the sense explained above under (ii).

It is quite remarkable that a purely affine connexion makes a comparison of lengths (which is a metrical concept) possible albeit only along a geodesic. Indeed, no natural comparison of length is afforded between different geodesics, even if they happen to cross each other, for the linear transformation of s alluded to above is free severally on every geodesic.

We ought to draw attention to the fact that according to (7.3) and (7.1) the skew part of $\Gamma^k{}_{lm}$ is irrelevant both for the geodesics and for the 'metric' on any geodesic, for it is obliterated by symmetry already in (7.1). Any skew tensor $\Theta^k{}_{lm}(= -\Theta^k{}_{ml})$ can be added to $\Gamma^k{}_{lm}$ without changing either.

The experience that any skew addition is irrelevant elicits the question: are there also symmetric additions to an affinity which do not change its system of geodesics? The answer is that an addition of the form

$$\delta^k{}_l V_m + \delta^k{}_m V_l$$

to $\Gamma^k{}_{lm}$ (where V_l is an arbitrary vector field) is the only symmetric addition not to change the geodesics of $\Gamma^k{}_{lm}$. It does, however, change the 'metric' on some of the (actually most of the) geodesics, however the field V_l be chosen. This includes that no change in the symmetric part of an affinity is possible, if both all the geodesics and the metric on all of them is to be preserved.

I leave it to the reader to prove the last statements.

CHAPTER VIII

THE GENERAL GEOMETRICAL HYPOTHESIS ABOUT GRAVITATION

THE UNDERLYING IDEA

It is far beyond the scope of these lectures to report on the development of the ideas, first of Restricted, then of General, Relativity and to show how they are logically built on the outcome of a number of crucial experiments, as the aberration of the light of fixed stars, the Michelson-Morley experiment, certain facts regarding the light from visual binary stars, the Eötvös-experiments which ascertained to a marvellously high degree of accuracy the universal character of the gravitational acceleration—that is to say that in a given field it is the same for any test-body of whatever material.

Yet before going into details about the metrical (or Riemannian) continuum, I wish to point out the main trend of thought that suggests choosing such a one as a model of space-time in order to account for gravitation in a purely geometrical way. In this I shall not follow the historical evolution of thought as it actually took place, but rather what it might have been, had the idea of affine connexion already been familiar to the physicist at that time. Actually the general idea of it emerged gradually (in the work of H. Weyl, A. S. Eddington and Einstein) from the special sample of an affinity that springs from a metrical (Riemannian) connexion—emerged only after the latter had gained the widest publicity by the great success of Einstein's 1915 theory. Today, however, it seems simpler and more natural to put the affine connexion, now we are familiar with it, in the foreground, and to arrive at a metric by a very simple specialization thereof.

We have learnt that in the particularly simple case of a symmetric *integrable* affinity a frame of coordinates can be found in which the geodesics are straight lines.† Moreover, we know from ordinary

† For in this case the Γ's can be transformed to zero everywhere, so that (7.3) defines straight lines. The restriction to *symmetrical* integrable affinities is required, even though the skew part does not affect the shape of the geodesics. For the symmetrical part of an integrable affinity need not (and as a rule will not) be integrable and therefore *cannot* be transformed away.

mechanics that the path of a particle, not acted on by any force, *is* a straight line, both in space and in space-time (since the motion is in this case uniform). Putting it more cautiously and much more significantly for our present purpose: it is a straight line in a suitably chosen frame of reference, the same for all particles not subject to a force, a so-called inertial frame. But the path would not be straight in space-time, that is, the spatial path would not be straight and the motion not uniform, when referred to a system of coordinates that has itself an accelerated or a rotational motion with respect to an inertial frame, as, for example, the spatial frame fixed rigidly to the rotating Earth has.

Now from Eötvös' experiments we infer that in a given field of gravitation any particle of whatever nature, when starting from a given point in space-time (i.e. from a given point in space at a given time) in a given direction in space-time (i.e. in a given direction in space with a given velocity), follows a curve (we shall call it a 'world-line') that depends only on the said initial conditions and on the gravitational field, not on the nature of the particle. Moreover, this curve is not a straight line when referred to an inertial frame. Or better, since it may be doubtful what an inertial frame means in this case because there are no particles exempt from gravitation: these curves are not straight lines in any frame; there is no frame in which they are all straight, with one not notable, because rather fictitious, exception.†

This state of affairs suggests tentatively extending the analogy between the geodesics of an integrable affinity and the paths (or 'world-lines') of particles not subject to any force to the geodesics of a general, non-integrable affinity and the paths of particles subject to the action of a gravitational field. The temptation is particularly strong, because the geodesic, by its definition, may patently be called the 'straightest' line, so that we would have the simple law: a particle follows in all cases the *straightest* line—a law that is not without precedent. It is strongly reminiscent of the well-known result of ordinary, classical mechanics, viz. that a particle constrained to remain on a given surface, but otherwise not subject to

† It is the case of a strictly uniform field of gravitation, which does not exist. Unfortunately this fictitious case has become the stock example in all popular or semi-popular treatments.

any force, moves with uniform velocity along a geodesic of this surface.

In other words, we assume that a gravitational field can be pictured as a purely geometrical property of space-time, namely as an affinity imposed upon it, and that it amounts to a geometrical constraint on the motion of particles. This affine connexion is to be regarded as an inherent property of the space-time continuum, not as something that is created only when there is a gravitational field. The case where there is none is simply the case where the affinity is integrable.

In these considerations we have tacitly adopted a very relevant generalization of the *classical* idea of a 'frame' which must not be passed over in silence on this occasion, though it has become familiar to us from the preceding chapters. It is not only that we include time in a quite general way in the transformation of coordinates. But a classical physicist, when speaking of the inertial frame or any other frame, had only in mind that the Cartesian coordinates of a point could be referred to either of two rigid systems of axes which move with respect to each other as a thrown stone moves with respect to the Earth or the Earth with respect to an inertial system. In this case the coordinates in one frame are special linear functions of those in another frame, with coefficients that are some functions of the time. We, however, have inadvertently switched over to contemplate our completely general transformations, which are linear only in the near vicinity of a point; and the coefficients (the $\partial x_i/\partial x'_k$) are *arbitrary* functions of all four coordinates and change from point to point. To justify this generalization we may say that without it the general idea of affine connexion would not come in at all and so could not be used to picture the gravitational field.

Another remark is useful. Having adopted this general idea of a frame of reference, we do not wish to grant a prerogative to any special frame. Hence whenever, in the particular frame we are using, the components of the affine connexion are not all zero everywhere, we are obliged to regard them as representing a gravitational field from the point of view of this frame, even though we might be inclined to call it a sham field in the case of an integrable affinity, when a frame can be found in which they all vanish. But if, on account of that, we disregarded them in the original frame where they do not vanish, we should not draw the geodesics correctly. Also, we do not want to make exceptions in that special case, we do not wish to impose the rule that we must always bother whether or not the affine connexion as a whole can be reduced to nothing and, if so, bother to adopt the frame where it is so reduced.

Perhaps it would be interesting to know whether at least in the neighbourhood of a particular point the gravitational field can be 'transformed away'. The answer to this is simple: it always can. But we will come to this later.

THE LAW OF GRAVITATION

In Newton's theory gravitation is described by a potential ϕ, the gradient of which, taken negative, is the acceleration imparted to a small test-body. The law governing the ϕ-field reads

$$\phi = \text{const., where there is no field;} \qquad (8.1)$$

$$\nabla^2\phi = 0, \qquad (8.2)$$

where there is a field, but no gravitating matter; and

$$\nabla^2\phi = 4\pi k\rho, \qquad (8.3)$$

where there is gravitating matter of density ρ, k being the constant of gravitation, 6.67×10^{-8} g.$^{-1}$ cm.^{3} sec.$^{-2}$.

Obviously and of necessity, (8.1) is included as a special case in (8.2), and the latter as a special case in (8.3). Or, putting it the other way round, (8.2) is a generalization of (8.1), and (8.3) a generalization of (8.2).

Since we wish to represent the field by an affine connexion Γ^i_{kl}, the cardinal question is, what are the corresponding laws governing the Γ^i_{kl} in these three cases?

There is no doubt about the analogue of (8.1). Where there is no field, the connexion must be integrable and the geodesics straight lines. We learnt in Chapter VI that the necessary and sufficient condition for this to be so is

$$B^i_{klm} = 0; \qquad \Gamma^i_{kl} = \Gamma^i_{lk}. \qquad (8.4)$$
$$\text{(80 equations)} \quad \text{(24 equations)}$$

So this must be fulfilled where there is no field (and no matter). The condition to impose on the Γ's where there is a field but no matter must be expressed by one (or more) tensor-equations, which must be fulfilled *inter alia* in the limiting case of no field, that is to say, they must be a mathematical consequence of (8.4), demanding, however, less than (8.4). That leaves us still with a wide choice, as an attempted generalization usually does. For example, the tensor-equations

$$B^i{}_{klm} B^k{}_{pqr} = 0,$$

$$B^i{}_{klm} B^k{}_{pqi} = 0,$$

$$R_{lq} R_{mr} + a B^i{}_{klm} B^k{}_{iqr} = 0, \quad (a = \text{some constant})$$

and many others fulfil the requirement. So let us be guided by the principle of simplicity, which suggests that it is worth while trying an equation or equations which are linear at least with respect to the derivatives of the Γ's (as equation (8.4) actually is) if such there are. (Notice that the classical equations (8.1)–(8.3) are altogether linear.) If we add this demand, then products of B's are excluded, and the only way† of deducing from (8.4) anything less exigent than (8.4) itself is to contract it. Thus we get

$$R_{kl} = 0, \quad (S_{lm} = 0). \tag{8.5}$$

Now remember, on the one hand, that if $\Gamma^i{}_{kl}$ is a symmetric affinity, the second equation is contained in the first and can be scratched; on the other hand, it seems very much worth while to try whether we can do with symmetric Γ's, because the skew part would have no influence whatever on the geodesics, which, after all, inspired our whole attempt. We do that, and henceforth *until further notice take the affinity to be symmetric*

$$\Gamma^i{}_{lk} = \Gamma^i{}_{kl}. \tag{8.6}$$

Then we are left with

$$R_{kl} = 0, \tag{8.7}$$

as the general equation to impose on the Γ's *where there is no matter.*

This is in one respect very satisfactory, in another respect not quite so. Let us first speak of the satisfactory point. It is, that the vanishing of a tensor of the second rank in empty space is just what we would expect as the mathematical description of the concept 'empty', i.e. devoid of matter. For, according to the famous

† Merely to drop the 24 conditions leads to Einstein's 'Fernparallelismus'. It did not work.

identification of mass and energy, which Einstein inferred from a simple thought-experiment on the pressure of light and which was so strikingly confirmed by actual experiments on the disintegration of matter by nuclear collisions that its fatal large-scale confirmation by the 'atomic bomb' was quite gratuitous—I say, according to this famous discovery of Einstein, matter is not represented by a scalar but by a tensor of the second rank, because energy is not a scalar; it is the time-time component of the stress-energy-momentum (or flux-energy-momentum) tensor.† Using the terminology of the Restricted Theory of Relativity, though we cannot explain it in detail at the moment, we have that for a particle of rest-mass m this tensor is

$$m\frac{dx_i}{ds}\frac{dx_k}{ds}, \tag{8.8}$$

where ds is the differential of proper time (an invariant).

So we may take it that our R_{kl} is essentially the matter-tensor, that (8.7) expresses the fact that it vanishes (*empty* space) and that the generalization of (8.7) inside matter, corresponding in Newton's theory to the transition from (8.2) to (8.3), will be of the form

$$R_{kl} = CT_{kl}, \tag{8.9}$$

where C is a constant and T_{kl} is the matter tensor. This interpretation will prove to be not yet quite correct; it will require a slight readjustment.

Pending a detailed investigation, this is satisfactory as far as it goes. It is slightly disturbing that the tensor (8.8) is contravariant, while in (8.9) a covariant tensor is required. It is much more disturbing that the tensor (8.8), as well as any elementary energy tensor, e.g. the Maxwellian, is symmetric, while R_{kl} is not, not even with our symmetric affinity—we had drawn attention to this fact in Chapter VI. But the most disconcerting fact is that there are 40 functions Γ^i_{kl}, which cannot be sufficiently controlled by the equations (8.7) or more generally speaking (8.9), which number only 16 and will have to be reduced to 10, since we will have to get rid of the asymmetry in R_{kl} anyhow.

† That the energy is not an invariant can be seen from the elementary consideration that the kinetic energy of a particle vanishes in some inertial frames but not in all of them.

Both the disturbing asymmetry in R_{kl} and the shortcoming of the number of equations seem to indicate that for representing a pure gravitational field something much less general than a symmetric Γ^i_{kl} with 40 independent components must be contemplated, in other words that some further general restriction must be imposed on the connexion. Now this necessity is supported from an entirely different side that will show us the way.

PART III

METRICALLY CONNECTED MANIFOLD

CHAPTER IX

METRICAL AFFINITIES

GENERAL INVESTIGATION

Two circumstances combine to let us think that with the basic affine connexion there must in some way be associated another geometric entity of fundamental significance, viz. a Riemannian metric. Actually it was from this side that Einstein first attacked the problem of the structure of space-time. The notion of affinity was brought in later by H. Weyl.†

The first circumstance is that, as we saw, an affine connexion already gives rise to an invariant ds along every geodesic. Comparison of 'length' or 'interval' (it is not really just length, remember we are in four dimensions) becomes possible.

The idea suggests itself that this comparison of intervals should not be restricted just to one and the same geodesic.

The second circumstance is that such an invariant ds is actually known in the so-called Restricted Theory of Relativity. We shall enter into the details later. It is *not* the sum of squares, but $dt^2 - d\sigma^2$ ($= ds^2$ say), where $d\sigma$ means the spatial element of distance. A generalization thereof is the general line-element that we will have to consider:

$$g_{ik}\, dx_i\, dx_k$$

(where it is sufficient to take $g_{ik} = g_{ki}$). What is the connexion between the two 'ds'? We will obviously have to demand that the primitive Γ-metric forms part of this g_{ik}-metric.

We turn to a more thorough investigation of these relationships. If x_1, x_2, x_3 are interpreted as spatial coordinates and x_4 as the time, the components of the velocity of a particle at the point x_k are

$$\frac{dx_1}{dx_4}, \quad \frac{dx_2}{dx_4}, \quad \frac{dx_3}{dx_4}, \tag{9.1}$$

where $x_k + dx_k$ is a neighbouring point on the world-line of the particle. The transformation formulae of the three quantities (9.1) are easily derived from (1.4), but they are extremely unwieldy,

† *Raum, Zeit, Materie* (Berlin, Springer, 1918).

namely linear but non-homogeneous, and fractional. Now since, after all, dx_k *is* a vector, it is reasonable to envisage instead of (9.1) a definite vector with components proportional to dx_k (not dx_k itself, because it is not a *definite* vector) from which the quantities (9.1), if ever desirable, can be obtained as quotients. Moreover, it is reasonable to demand that such a vector should be always available.

For this purpose we need an infinitesimal invariant proportional to the dx_k. In the chapter on geodesics we have learnt that the affine connexion itself procures such an invariant, viz. the differential ds of the parameter s that is distinguished on every geodesic in that it gives to its equation the simple form (7.3). And we have learnt that the vector dx_k/ds is parallel-transferred along the geodesic. Yet it is again not a quite *definite* vector, because the distinguished variable is not quite unique, it is only determined up to a linear transformation with arbitrary constant coefficients ($s' = as+b$). Thus ds is only determined up to a constant multiplier (a) and so is dx_k/ds. This multiplier is still free on every one of the ∞^6 geodesics.

Can this lack of definiteness be removed, so that dx_k/ds becomes a definite vector on every geodesic and thus for every line-element? In principle that seems easy: just take an arbitrary definite choice of s, independently on every geodesic.

Well, we shall see how that works. *After we have taken our choice*, ds will for every line-element dx_k be a definite homogeneous invariant function of the first degree of the dx_k. We shall not set to explore all the vast possibilities which that leaves, but only the one suggested by the elementary way distance is measured in a skew Cartesian system of coordinates—essentially by the Pythagorean theorem. That is, we assume

$$ds^2 = g_{ik}\,dx_i\,dx_k, \tag{9.2}$$

where g_{ik} is a symmetrical tensor, varying from point to point. This very special assumption is reasonably justified by the apprehension that no other one would make our model join to the more elementary concepts of physics. Yet we must be aware that we thus impose a considerable restriction which is not likely to be compatible with an *arbitrary* affinity.

We wish to know the necessary and sufficient condition for (9.2) to be in accordance with the affine measure of distance along every geodesic. The answer we have to expect is a relationship between the tensor g_{ik} and the connexion Γ^k_{lm}. The task is not quite easy. We shall first explore at some length a *sufficient* condition, which is *not* necessary, but will lead us by itself to the less restrictive *necessary and sufficient* condition.

I maintain, a *sufficient* condition is that the invariant

$$g_{ik}A^iA^k, \tag{9.3}$$

where A^k is any vector (not vector-field) at any point, be conserved on any parallel displacement of the vector A^k.

Indeed let $\hat{s}$ (to distinguish it for the moment from the s in (9.2)) be the affine parameter chosen on a given geodesic. The invariant

$$g_{ik}\frac{dx_i}{d\hat{s}}\frac{dx_k}{d\hat{s}} = C \quad \text{(say)}$$

will then be conserved on parallel transfer of $dx_i/d\hat{s}$ along this geodesic. Hence if you replace, as you may, the parameter $\hat{s}$ on this geodesic by $\hat{s}\sqrt{C}$ and call that s, (9.2) is fulfilled. Since you can do the same on every geodesic, *the sufficiency is proved.*

Since an invariant product is certainly conserved on parallel-transfer when *all* its factors are parallel-transferred (see Chapter v), the condition is certainly fulfilled if the affinity Γ^i_{kl} transfers the field g_{ik} into itself, that is, if the invariant derivative of g_{ik}, taken with respect to Γ^i_{kl}, vanishes:

$$g_{ik;l} \equiv \frac{\partial g_{ik}}{\partial x_l} - g_{mk}\Gamma^m_{il} - g_{im}\Gamma^m_{kl} = 0. \tag{9.4}$$

Moreover, you easily realize, that none but the parallel-transferred g_{ik} can conserve the invariant (9.3) for an arbitrary A^k. Indeed with A^k arbitrary its parallel-transferred is also arbitrary. Hence the (transferred) invariant is known for an arbitrary (transferred) vector, and by this the g_{ik} in the new place is determined uniquely.

Thus (9.4) is the mathematical expression of our *sufficient* condition. We write the three equations that result from a cyclic permutation of the subscripts ikl:

$$\begin{array}{l|c}
\dfrac{\partial g_{ik}}{\partial x_l} - g_{mk}\Gamma^m_{il} - g_{im}\Gamma^m_{kl} = 0 & -\tfrac{1}{2} \\[2ex]
\dfrac{\partial g_{kl}}{\partial x_i} - g_{ml}\Gamma^m_{ki} - g_{km}\Gamma^m_{li} = 0 & +\tfrac{1}{2} \\[2ex]
\dfrac{\partial g_{li}}{\partial x_k} - g_{mi}\Gamma^m_{lk} - g_{lm}\Gamma^m_{ik} = 0 & +\tfrac{1}{2} \\
\hline
\end{array}$$

and combine them with the factors indicated beyond the bar. In doing so we take into account the symmetry of g_{ik}, but not that of $\Gamma^m{}_{kl}$. In other words, for the moment we proceed to determine the most general non-symmetric affinity that complies with our sufficient condition. (The reason will appear later. We are not really out for the non-symmetric affinities. But this procedure will facilitate the finding of the sufficient *and* necessary condition.) We get:

$$\frac{1}{2}\left(\frac{\partial g_{kl}}{\partial x_i}+\frac{\partial g_{li}}{\partial x_k}-\frac{\partial g_{ik}}{\partial x_l}\right)-\tfrac{1}{2}g_{lm}(\Gamma^m{}_{ki}+\Gamma^m{}_{ik})$$
$$+\tfrac{1}{2}g_{im}(\Gamma^m{}_{kl}-\Gamma^m{}_{lk})+\tfrac{1}{2}g_{km}(\Gamma^m{}_{il}-\Gamma^m{}_{li})=0. \quad (9.5)$$

We can solve these equations with respect to the *symmetric* part of Γ, with the help of the tensor g^{ik} derived from the tensor g_{ik} in the way described in Chapter II (see equations (2.8)–(2.10)) and uniquely determined by

$$g^{ik}g_{lk}=\delta^i{}_l.$$

In the present case it is obviously also symmetric.

We 'multiply' (9.5) by g^{sl} and put for abbreviation

$$\left.\begin{aligned}\tfrac{1}{2}(\Gamma^m{}_{ik}+\Gamma^m{}_{ki}) &= \Gamma^m{}_{\underline{ik}},\\ \tfrac{1}{2}(\Gamma^m{}_{ik}-\Gamma^m{}_{ki}) &= \Gamma^m{}_{\check{ik}}.\end{aligned}\right\} \quad (9.6)$$

$$\tfrac{1}{2}g^{sl}\left(\frac{\partial g_{kl}}{\partial x_i}+\frac{\partial g_{li}}{\partial x_k}-\frac{\partial g_{ik}}{\partial x_l}\right)=\begin{Bmatrix}s\\ i\,k\end{Bmatrix}. \quad (9.7)†$$

We obtain (adding $\Gamma^s{}_{\check{ik}}$ on both sides):

$$\Gamma^s{}_{ik}=\begin{Bmatrix}s\\ i\,k\end{Bmatrix}+g^{sl}g_{im}\Gamma^m{}_{\check{kl}}+g^{sl}g_{km}\Gamma^m{}_{\check{il}}+\Gamma^s{}_{\check{ik}}. \quad (9.8)$$

This formula gives the complete answer to the question which affinities transfer a given g_{ik}-field into itself. Observe that both the curly bracket and the sum of the second and third terms on the right are symmetric in i and k. Hence the skew part $\Gamma^s{}_{\check{ik}}$ can be chosen arbitrarily, the even part $\Gamma^s{}_{\underline{ik}}$ (viz. the first three terms on the right) is then uniquely determined by the skew part and the g_{ik}-field. From (9.8) follows that the Christoffel-brackets

$$\Gamma^s{}_{ik}=\begin{Bmatrix}s\\ i\,k\end{Bmatrix} \quad (9.9)$$

† These are called the Christoffel brackets. They do not constitute a tensor, but according to (9.9) an affinity.

form the only *symmetric* connexion that complies with (9.4) for given g_{ik}-field. But we learn from (9.8) something more, even if we are interested in symmetric connexions only. We have seen in Chapter VII that neither the geodesics nor their 'affine-metrical' parameter depends on the skew part of the affinity. So let us just scrap it; then we are left with

$$\Gamma^s_{ik} = \begin{Bmatrix} s \\ i\,k \end{Bmatrix} + g^{sl} g_{im} \Gamma^m_{\check{kl}} + g^{sl} g_{km} \Gamma^m_{\check{il}}. \qquad (9.10)$$

This family of *symmetric* affinities, in which $\Gamma^m_{\check{kl}}$ is an arbitrary skew tensor, is equally well 'compatible' with the metric g_{ik}, though it has, of course, other geodesics than (9.9) and does not comply with (9.4), showing that the latter condition, while sufficient, is *not* necessary. It is now easy to derive the necessary and sufficient condition.

On the one hand the tensor, added on to the Christoffel brackets in (9.10), may be written—or rather (9.10) may be written

$$\Gamma^s_{ik} = \begin{Bmatrix} s \\ i\,k \end{Bmatrix} + g^{sl} T_{lik}, \qquad (9.11)$$

$$T_{lik} = T_{lki}, \qquad (9.12)$$

and it is easy to satisfy oneself

(i) that the T-tensor in addition to its symmetry in i and k fulfils the peculiar symmetry-condition

$$T_{lik} + T_{ikl} + T_{kli} = 0; \qquad (9.12a)$$

(ii) that it is otherwise arbitrary, since $\Gamma^m_{\check{kl}}$ is so. Indeed by taking

$$\Gamma^m_{\check{kl}} = -\tfrac{1}{3} g^{ms} (T_{ksl} - T_{lsk})$$

in (9.10) and observing (9.12) and (9.12a), you obtain (9.11). The last three relations are thus equivalent to (9.10), and the whole family of affinities described by them is what we may call 'compatible' with the metric g_{ik}. (Let this be a short expression for—not a hard and fast postulate of—the kind of agreement between them which we spoke of in detail.)

In reviewing our findings concerning this compatibility we shall now speak of symmetrical affinities only, to which, as we know, any skew part may be added without changing the geodesics and their affine metric, thus without interfering with compatibility. The *sufficient* compatibility condition (9.4) singles out (9.9) as the only symmetric affinity that fulfils it with a given g_{ik}; but we have just now found a whole class of symmetric affinities compatible with a given g_{ik}, namely (9.11) cum (9.12) and (9.12*a*). We proceed to show that this is already the widest class, in other words that the last three relations together represent the sufficient and necessary conditions of compatibility.

Observe first that even with a fixed g_{ik} only (9.12*a*) represents any restriction at all on the symmetric affinity (9.11), since the difference of two symmetric affinities, the Γ- and the Christoffel affinity, is a symmetric tensor anyhow, and that is just what the equations (9.11) and (9.12) say about it. Therefore it only remains to be shown that (9.12*a*), in addition to being sufficient, is also *necessary*.

Compatibility demands that the invariant (9.3) should be unity, if the *affine* dx_k/ds is taken for A^k. This must hold everywhere along a geodesic, and since here the direction-vector is parallel-displaced, the invariant (9.3) must not change, when the vector A^k is parallel-displaced according to the affinity (9.11) along a line-element ηA^k (with η an infinitesimal constant), while the g_{ik} change to their values in the neighbouring point. Now this latter change is known to cancel exactly those terms that would originate from displacing A^k by the Christoffel affinity alone. Hence in the whole operation the majority of terms are known to cancel; only those containing the tensor T survive and must vanish by themselves. This leaves us with the following conditions, which must be imposed on the components of the tensor T:

$$0 = -2g_{ik}A^i g^{ks} T_{slm} A^l A^m \eta,$$

that is to say
$$0 = -2\eta A^i A^l A^m T_{ilm}.$$

Now remember that A^k is arbitrary. By taking first only one of its components different from zero, then two of them, finally three of them, and by using (9.12) you easily prove (9.12*a*); the latter represent twenty independent conditions in addition to the former, and reduce T to twenty independent components.

This completes the proof, that (9.11)–(9.12*a*) are the necessary and sufficient conditions for a symmetric affinity Γ^i_{kl} and a metrical tensor g_{ik} to agree in the sense that the complete g_{ik}-metric accords with the incomplete Γ-metric, defined only along each affine geodesic. And, to repeat this, an arbitrary skew-symmetric tensor, added to our Γ, interferes with nothing, because it changes neither the affine geodesics nor their affine metric.

On how many independent functions does our general 'metric affinity' (as we may suitably call it) depend? There are 10 independent g_{ik}. The tensor T_{klm} is restricted to 40 independent components by the symmetry in l and m. A careful count shows that (9.12*a*) amounts to 20 independent conditions. Our tensor has thus 20 independent components and our Γ seems therefore to depend on 30 arbitrary functions.

At the end of the previous section the desire was felt to restrict Γ so as to depend on 10 functions only. That suggests envisaging after all only the simplest case, viz. $T_{klm} = 0$. This ruling also meets another desire that was felt, viz. to get the Einstein-tensor R_{kl} symmetric. With a symmetric Γ the only term that disturbs the symmetry is, from (6.17)

$$\frac{\partial \Gamma^{\alpha}{}_{k\alpha}}{\partial x_l}. \tag{9.13}$$

We shall see presently that for the Christoffel affinity (9.9) it is indeed symmetric in k and l.

By accepting (9.9) and thereby (9.4) we have now reached exactly the geometrical point of view underlying Einstein's 1915 theory, known as the General Theory of Relativity, which is going to occupy us for several chapters. Even before entering into any details about it, we have become aware that it represents, from the affine standpoint, a very special case, capable of generalization in more than one direction.

SOME IMPORTANT FACTS AND RELATIONS

Let us right away become familiar with a few simple facts and conventions, all turning on the fact (9.4), that the invariant derivative of the metrical tensor g_{ik}—often called the fundamental tensor of this theory—vanishes.

First, it follows that also

$$g^{ik}{}_{;l} = 0.$$

For, from

$$g_{ik}g^{il} = \delta_k^l$$

by invariant differentiation:

$$g_{ik;m}g^{il}+g_{ik}g^{il}{}_{;m} = \delta_k{}^l{}_{;m} = 0.$$

Multiplying this by g^{ks}, you find

$$g^{sl}{}_{;m} = -g^{ks}g^{il}g_{ik;m} = 0.$$

As stated, a fundamental scalar density is the square-root of the determinant $\sqrt{g}$. What about its semicolon derivative? A determinant is a polynomial of all its $n^2(=16)$ components. Differentiating it with respect to every single component g_{ik} and remembering that the 'co-factor' is gg^{ik}, we get

$$\frac{\partial g}{\partial x_l} = gg^{ik}\frac{\partial g_{ik}}{\partial x_l}.$$

We replace the derivative on the right by its value drawn from (9.4):

$$\frac{\partial g}{\partial x_l} = gg^{ik}(g_{mk}\,\Gamma^m{}_{il}+g_{im}\,\Gamma^m{}_{kl})$$

$$= 2g\Gamma^\alpha{}_{\alpha l};$$

for this we can write $$\frac{\partial\sqrt{g}}{\partial x_l}-\sqrt{g}\,\Gamma^\alpha{}_{\alpha l} = 0$$

or $$\sqrt{g}_{;l} = 0.$$

Incidentally we have now supplied the proof that the metrical Einstein-tensor *is* symmetric. For the ostensibly disturbing term (9.13) may now be written

$$\frac{\partial\Gamma^\alpha{}_{k\alpha}}{\partial x_l} = \frac{\partial^2\log\sqrt{g}}{\partial x_l\,\partial x_k},$$

which is symmetric. Summarizing g_{ik}, g^{ik}, $\sqrt{g}$ have vanishing semicolon derivatives.

This fact makes a certain convention about 'drawing' (or 'raising' and 'lowering') of indices a particularly convenient tool. We associate with any tensor of a definite description,

$$T^{klm\ldots}{}_{pq\ldots},$$

many (to be precise, 2^s-1, if s is its rank) others, which differ from it in character by one or more of the superscripts being 'lowered' and/or one or more of the subscripts being 'raised'. The procedure for obtaining these tensors is illustrated by

$$T_k{}^{lm\ldots}{}_{pq\ldots} = g_{ks}T^{slm\ldots}{}_{pq\ldots},$$

$$T^{klm\ldots}{}_p{}^q{}_{\ldots} = g^{qs}T^{klm\ldots}{}_{ps\ldots}.$$

Raising and subsequent lowering of the same index (or vice versa) leads back to the original tensor. It is immediately clear that, with this convention introduced, one has to establish and carefully to observe a definite order between *all* indices, not only between those of the same character as before, since they can change their character. One has taken to the habit of regarding all these different associated tensors as the *same* tensor. This is quite legitimate, but of course it refers to a definite fundamental tensor g_{ik} given once and for all.

In very much the same way one can also associate with any tensor a tensor density and vice versa, by multiplying or dividing it by $\sqrt{-g}$. (Apart from rare and odd exceptions at isolated points, g is assumed to be always different from zero and negative.)

It is easy to prove, but it deserves special mention, that a couple of *dummy* indices can be raised and lowered simultaneously without any further effect. For instance, we have identically (according to our conventions)

$$T^{klm\ldots}{}_{kq\ldots} \equiv T_k{}^{lm\ldots k}{}_{q\ldots}.$$

Now these conventions are particularly convenient on account of the fact that the fundamental tensors g_{ik}, g^{ik}, $\sqrt{g}$ are their own parallel-transfers or, what is the same, have zero derivatives.

From the first way of expressing this fact it follows immediately that the associations in question are conserved on parallel displacement. Of even greater practical use is the consequence that drawing of indices can be effected 'inside the semicolon'. It will suffice to make the point clear by an example. Say you are given the equation

$$B^k{}_{;i} = t^k{}_i.$$

Then you can infer that $B_{k;i} = t_{ki}$.

The reason is that

$$B_{k;i} = (g_{kl}B^l)_{;i} = g_{kl;i}B^l + g_{kl}B^l{}_{;i} = g_{kl}B^l{}_{;i}.$$

Similarly and for the same reasons 'latinizing' and 'gothicizing', i.e. dividing or multiplying by $\sqrt{g}$, may be effected under the semicolon.

It deserves to be mentioned that g^{ik}, though it was originally defined in a different way, actually is the fundamental tensor g_{ik}, with both its subscripts raised, exactly after our convention. If only one index is raised

$$g^{is}g_{sk} = \delta^i{}_k,$$

the result is the mixed unity tensor, *independent of the special metric*.

In contrast to what has just been said about the relationship between g_{ik} and g^{ik}, the following is remarkable. One sometimes has to contemplate an arbitrary *variation* of the fundamental tensor g_{ik}, say δg_{ik}. This entails automatically corresponding increments δg^{ik}. Now, if you vary in this way the preceding equation you get

$$\delta g^{is} g_{sk} + g^{is} \delta g_{sk} = 0,$$

or, multiplying by g^{kl},

$$\delta g^{il} = -g^{kl} g^{is} \delta g_{sk}.$$

One might have expected the + sign. Anyhow, it turns out that not δg^{ik} and δg_{ik}, but $-\delta g^{ik}$ and δg_{ik} are associated.

GEODESIC COORDINATES

We have in Chapter VI proved that a frame always exists in which all the components of a given *symmetric* affinity vanish at a given point. We call this geodesic coordinates at that point.

From (9.4) this means, in the present case, that all the derivatives $g_{ik,l}$ vanish at that point or that the g_{ik} are stationary there. It is sometimes very convenient to specialize for a moment in a geodesic frame, because some frame-independent relations may by this great simplification be discoverable at a glance, while they are not so patent in the general frame. To give an example, we envisage the Riemann-Christoffel-tensor of the Christoffel-bracket-affinity:

$$B^i{}_{klm} = -\begin{Bmatrix} i \\ kl \end{Bmatrix}_{,m} + \begin{Bmatrix} i \\ km \end{Bmatrix}_{,l} + \begin{Bmatrix} i \\ \alpha l \end{Bmatrix}\begin{Bmatrix} \alpha \\ km \end{Bmatrix} - \begin{Bmatrix} i \\ \alpha m \end{Bmatrix}\begin{Bmatrix} \alpha \\ kl \end{Bmatrix}.$$

From Chapter VI we are acquainted with some of the symmetry properties of the general R.Ch.-tensor. It is *always* antisymmetric in its last pair of subscripts and enjoys for any symmetric affinity the cyclic symmetry expressed in (6.16*a*). Let us now replace the brackets according to (9.7) and use a geodesic frame, which makes all *first* derivatives vanish. One obtains

$$B^i{}_{klm} = \tfrac{1}{2} g^{is}(-g_{ls,k,m} + g_{kl,s,m} + g_{ms,k,l} - g_{km,l,s}).$$

The associated totally covariant tensor

$$B_{sklm} = \tfrac{1}{2}(-g_{ls,k,m} + g_{kl,s,m} + g_{ms,k,l} - g_{km,l,s})$$

now exhibits two further symmetry properties, in addition to those mentioned just before. Namely, in addition to

(i) being antisymmetric in the second pair (l, m),

(ii) and having according to (6.16 a)

$$B_{sklm} + B_{slmk} + B_{smkl} = 0,$$

it is also

(iii) antisymmetric in the first pair (s, k), and

(iv) symmetric with regard to the exchange of the inner couple (k, l) among them, accompanied by an exchange of the outer couple (s, m) among them; or in other words (with a view to (i) and (iii)) an exchange of the first and second subscripts *with* the third and fourth.

No other symmetries independent of these can be found, *and hence there are none*. For in the simplified form they cannot be concealed, they would have to shew up, since the four second derivatives that appear are quite obviously capable of any arbitrary values. This negative conclusion is easily as important as the previous positive ones, though text-books usually fail to emphasize it.

Let us count the number of independent components of our tensor. By (i) and (iii) we must have $s \neq k$ and $l \neq m$, if the component is not to vanish. Since there are 6 such pairs of numbers, we first get $6 \times 6 = 36$ components. Among them are just 6 where the two pairs are identical, so that (iv) becomes trivial for them. For the remaining 30 components it is not trivial and reduces their number virtually to 15. So we are left now with $15 + 6 = 21$ components. Careful consideration of the only remaining cyclical property (ii) shows that it reduces this number just by one, *leaving us with* 20 *independent components*.

CHAPTER X

THE MEANING OF THE METRIC ACCORDING TO THE SPECIAL THEORY OF RELATIVITY

Our geometrical construction—a four-dimensional continuum with affine and metrical connexion—is to serve as a model of the real physical world. What physical interpretation are we to give to the 'line-element' ds—the infinitesimal invariant determined by every pair of infinitely neighbouring points x_k and $x_k + dx_k$?

In the beginning of Chapter IX the request for the invariant ds was prompted by the formal desire to obtain for the important notion of three-dimensional velocity a handier representation amenable to tensor calculus. Its elementary definition by the components

$$\frac{dx_1}{dx_4}, \quad \frac{dx_2}{dx_4}, \quad \frac{dx_3}{dx_4}$$

is unwieldy, since this is not a tensor. It does not vanish in every frame if it vanishes in one frame, and has pretty complicated, viz. *fractional*, linear transformation formulae.

Clearly ds must be related to some kind of 'distance' between the two points, which, however, be it remembered, are not two points in space, but two world-points, i.e. two neighbouring points in space, envisaged at two infinitely neighbouring moments in time (dx_4). According to a well-known algebraic theorem the original metrical definition of ds, viz.

$$ds^2 = g_{ik}\, dx_i\, dx_k \tag{10.1}$$

can, by a linear transformation of the dx_k with constant coefficients and a non-vanishing determinant,

$$dx_k = a_k^i\, dx'_i, \tag{10.2}$$

always be turned into an aggregate of squares only

$$ds^2 = \sum_{k=1}^{4} (\pm)\, dx_k'^2. \tag{10.3}$$

And, of course, a general coordinate transformation can always be indicated which produces this form at any particular point. (E.g. at the point $x_k = 0$, take the transformation so that *at that point* it has the development $x_k = a_k{}^i x'_i +$ higher powers of the x'_k.) From Euler's famous theorem on the 'inertia of quadratic forms' it is known that if the coefficients $a_k{}^i$ in (10.2) are confined to reality, the number of (–)-signs in (10.3) is not at our choice, it is invariably determined by the original coefficients g_{ik}. We can say more. Remember that the determinant of the g_{ik}, which we have called g, is on transformation multiplied by the square of the determinant of the a_k^i. So if these are real, the *parity* of the number of (–)-signs in (10.3) is determined by the sign of g, the number of (–)-signs being even for $g > 0$ and odd for $g < 0$. Now, since we have excluded $g = 0$ (save, quite exceptionally, at some *isolated* points), the sign of g can never change and therefore the parity in question, and therefore the exact number of (–)-signs can never change, because it could obviously do so only where $g = 0$ (isolated points do not matter, they can be avoided). So the number of (–)-signs is the same in the whole world and it is a matter of importance to make the appropriate choice for our model once and for all.

The clue to this number and to the meaning of ds is given by the Special Theory of Relativity. This theory starts from some everyday system of Cartesian space-coordinates (an inertial system, to be quite definite, that is to say one for which the ordinary laws of mechanics hold at least in the limit of small velocities of the moving bodies) plus a linear time-parameter, read from a good old grandfather's clock. The theory then contemplates a group of certain homogeneous, linear transformations with constant coefficients, transformations which involve all *four* coordinates and which there is reason to interpret as: going over to another inertial system, that moves with constant translational velocity with respect to the first. In calling it again an inertial system, we make the (well-founded) assumption that in it too the ordinary laws of mechanics hold in the limit of velocities that are small *in it*. (That, by the way, is the manner in which the theory extends the ordinary laws of mechanics to large velocities. For the relative velocity of the two spatial frames need not be small, though it is, by the nature of the transformation, limited to be smaller than the velocity of light.) In fact the very backbone of the theory is that all laws of Nature shall be the same for every frame reached in this way, including, of course, the original one from which we started; there should be no difference or distinction in principle between all these inertial frames, any one of which can be reached from any other one by a transformation of that group—a so-called Lorentz-transformation. In short, all laws of Nature are assumed to be invariant to Lorentz-transformation.

Now, since these linear transformations of special relativity (just as our more general ones) involve all four 'coordinates' x_1, x_2, x_3, x_4, you can, of course, identify the same *world-point* after the transformation, but there is no good meaning (just as little as in our more general case) in speaking of the same point in space after the transformation unless you also refer to the moment of time in which it is contemplated; neither is there a meaning in speaking of the same moment of time after the transformation without reference to the point in space where it is contemplated. What in one frame is the same point in space, envisaged at different moments of time, will in general turn out to be two different points in space in the other frame, envisaged at two different moments. Again, what in one frame is the same moment at two different points, will in general be mapped in the other frame as different moments referring to different points in space. It is this state of affairs which has given birth to all the much discussed 'paradoxes' in the Special Theory of Relativity—so difficult to explain to the non-mathematician, while the mathematician is prepared to encounter some clashes with customary views from the mere fact that all four coordinates are involved in the transformation.

From what has been said it is to be anticipated that neither the *distance* between two points in space (say, the distance of two points at which two well-defined momentary *events* happen) nor the *time interval* between the happening of the two events are invariant to Lorentz-transformation; either of them may even vanish in one frame, but not vanish in another frame. If we take for convenience one of the two events to happen at the origin at time zero, the other one at the point x_1, x_2, x_3 at time x_4, the square of their distance in that frame will be given by the Pythagorean theorem, thus

$$x_1^2+x_2^2+x_3^2$$

and their time interval by $\qquad x_4.$

Since all frames are to be of equal right, the same expressions will hold in any other frame, only with the x_k''s for the x_k. But neither is invariant. We shall have in general

$$x_1'^2+x_2'^2+x_3'^2 \neq x_1^2+x_2^2+x_3^2,$$

$$x_4' \neq x_4.$$

However—and this is the cardinal point—the Lorentz-transformation is characterized by the fact that the following expression (which could equally well be quoted with the opposite sign) *is* invariant

$$-x_1^2-x_2^2-x_3^2+x_4^2 = -x_1'^2-x_2'^2-x_3'^2+x_4'^2. \qquad (10.4)$$

I said the transformations are characterized by this invariance. Indeed it is well-nigh an exhaustive definition distinguishing Lorentz-transformations among all possible homogeneous linear transformations of the four coordinates. The state of affairs bears formal analogy to the case of orthogonal transformations (rotations of the Cartesian system) in three dimensions, which are characterized among all linear transformations by the invariance of the distance $x_1^2+x_2^2+x_3^2$.

I have still to comment on the 'well-nigh'. First, among the group of transformations delimited by the invariance (10.4) are such with transformation determinant $+1$ and such with -1. One usually excludes the latter on the ground that they cannot be reached from the 'identical transformation' ($x_k' = x_k$) by continuous change of the coefficients. Moreover, one usually demands that the coefficient which gives $\partial x_4'/\partial x_4$ shall be positive, for obvious reasons: one does not wish the time to increase in opposite directions in different frames.

A formal description of the injunction (10.4) can be given in the following terms. If you write down the four linear transformation formulae and transcribe every term containing x_4 or x_4' thus:

$$ax_4 = (-ia)(ix_4), \quad x_4' = (-i)(ix_4'),$$

in other words if you regard it as a transformation between the variables x_1, x_2, x_3, ix_4, and x_1', x_2', x_3', ix_4', then this is an *orthogonal* transformation (or a 'rotation') in four dimensions but with some relevant injunctions on the coefficients as to being either real or purely imaginary.

The invariant (10.4) of two world-points or 'point-events', one of which was for simplicity taken to happen at the origin of the four-dimensional frame, is the square of the time-interval minus the square of the distance. It can be positive or negative. In the first case x_4' can never vanish, and thus, in virtue of our conventions about the determinant and the coefficient $\partial x_4'/\partial x_4$, it cannot change its sign on Lorentz-transformation. The second event (the one with coordinates x_k) is then called *later* or *earlier* with respect to the first (the one at the four-dimensional origin) according to whether $x_4 \gtrless 0$. The term '*eigentime*' (more generally, eigentime-interval between the two events) has been adopted for the absolute value of the square root of the invariant (10.4) in this case. There is namely in this case a frame (to be reached by a particular Lorentz-transformation) in which the distance in space between the two events vanishes, in other words in which the two events are considered to happen at the same place. It is *that* frame in which a material point that in the original frame moves straight with uniform velocity in the time interval $0 \rightarrow x_4$ from the space point (0, 0, 0) to the space point (x_1, x_2, x_3) is considered to be at rest all the time. The eigentime is the time-interval between the two events which a 'Lilliput observer' moving in this way would read off his Lilliput stop-watch. In view of the invariance of (10.4) it is the shortest time-interval recorded between them in *any* frame.

We turn to the case when the invariant is negative. In this case the time-interval x_4' can vanish and change sign on Lorentz-transformation in spite of our convention about the determinant and the coefficient $\partial x_4'/\partial x_4$. The time order between the two events is not settled in an invariant way. The meaning of the invariant is: negative square of their distance in a frame in which the two events are simultaneous. It is the smallest distance they can acquire in any frame. I am not aware that the absolute value of the square root of the invariant has received in this case a name nearly as popular as the term of eigentime in the previous case. One might call it the distance of simultaneity or, shorter, the simultaneous distance. But these are not established terms. A slightly more involved notion has been given prominence. If a straight rod of suitable length is placed with its ends at space-points (0, 0, 0) and (x_1, x_2, x_3) and kept at rest in the original frame, you can find out its length, viz. $\sqrt{(x_1^2+x_2^2+x_3^2)}$, by inspecting the space coordinates of its extremities, and it is immaterial whether you do this simultaneously or not, since the rod is at rest. We choose to do it at the times 0 and x_4 respectively, where x_4 is selected so that in a certain other frame, with respect to which the rod is in uniform translational motion, $x_4' = 0$. In this frame it is essential that the two inspections should be simultaneous in it, if $\sqrt{(x_1'^2+x_2'^2+x_3'^2)}$ is to mean the length of the rod *in it*. Our invariant, applied to these two events (the two inspections), gives

$$-x_1^2-x_2^2-x_3^2+x_4^2 = -x_1'^2-x_2'^2-x_3'^2.$$

Hence in general

$$\sqrt{(x_1^2+x_2^2+x_3^2)} > \sqrt{(x_1'^2+x_2'^2+x_3'^2)}.$$

The rod has maximum length in the frame in which it is at rest. This is called its rest-length (German: *Ruhlänge*). The foreshortening in another frame (or also: when having been set in motion) is the famous Lorentz-Fitzgerald contraction. Though intimately connected with what I called 'simultaneous distance' of two given point-events with negative invariant, the notions of rest-length and Lorentz-contraction are, as I said, slightly more involved. For the concept of length of a rod does not refer to a pair of given point-events, the same in every frame. It refers to a pair of point-events of which one at least changes from frame to frame; viz. to two inspections (simultaneous in *that* frame) of the spatial coordinates of the two extremities of the rod.

We continue to investigate the two events (or world-points) with coordinates 0, 0, 0, 0 and x_1, x_2, x_3, x_4 respectively. We have still to consider the limiting case when their invariant (10.4) is neither positive nor negative, but vanishes. In this case neither can the time-interval be reduced to zero with the distance remaining finite nor vice versa. That both should be made zero simultaneously would not militate against the invariance of (10.4), but against the fact that the transformation is a one-to-one correspondence of world-points. On the other hand, by regarding this case as the limit of either of the other two, we anticipate the actual situation: there is now no lower limit, neither for the time-interval nor for the distance in space. By a suitable Lorentz-transformation both can be made as small as we choose—and that simultaneously, since they must always remain equal—without, however, *reaching* zero. (By a suitable Lorentz-transformation the time between emission and absorption of a quantum can be made as small as you choose.)

Hence now, as in the first case and unlike the second case, the *sign* of x_4 is invariant and we know which event is earlier. All the point-events which bear to the origin (0, 0, 0, 0) the relation of vanishing invariant lie on the manifold

$$-x_1^2-x_2^2-x_3^2+x_4^2 = 0 \qquad (10.5)$$

which is the (three-dimensional) surface of a 'spherical' (in one dimension less we would say 'circular') hyper-cone with its apex at the origin. It has literally the same equation in every frame. *That* half on which $x_4 > 0$ is called the cone of future (in German: '*Nachkegel*'), the other the cone of past ('*Vorkegel*'); all that with respect to the origin, which however represents any world-point and was chosen as one of our two point-events only for convenience.

The physical characteristic of any point-event on the cone of future is that it coincides in space and time with the arrival of a light-signal emitted from the origin, i.e. from space-point (0, 0, 0) at time zero. The physical characteristic of any point-event on the cone of past is that the origin (0, 0, 0, 0) lies on *its* cone of future; in other words that the 'origin-event' is simultaneous with the arrival of a light-signal emitted 'from that point-event', supposing it consisted in the emission of a light-signal.

These are simple consequences of (1) the earlier-later relation, depending on $x_4 \lesseqgtr 0$; (2) the equality of time-interval and distance; (3) a remark which we might have made earlier, namely that all our statements have been tacitly simplified by assuming that our co-ordinates shall measure length and time in such units as to make the velocity of light equal to 1. For instance, when using for time the second, we have to measure length in 'light seconds' (1 light second $= 3 \times 10^{10}$ cm.). Or, and that is usually preferred, if we keep to the cm. for length, we must use for time the 'light centimetre', the time light takes to cover the distance of 1 cm., i.e. $\frac{1}{3} \times 10^{-10}$ sec. (I believe some authors call the latter unit a light second. But that is in flat contradiction with what we call a light year, which is a measure of distance, not of time.) The light-cone neatly separates the region where our invariant (10.4) is positive or where

$$\sqrt{(x_1^2 + x_2^2 + x_3^2)} < |x_4| \quad \text{(interior of the light-cone)}$$

from the one where it is negative or where

$$\sqrt{(x_1^2 + x_2^2 + x_3^2)} > |x_4| \quad \text{(exterior of the light-cone).}$$

The latter is also called the region of simultaneity or of virtual simultaneity (all that with respect to our four-dimensional origin). The terminology is clear from our previous discussion. A direction pointing from the origin into the interior is called time-like; when pointing to a point in the region of simultaneity space-like, because the former can be chosen as x_4-axis in a Lorentz-frame, the latter, for example, as x_1-axis. A signal of any kind (including a moving particle, a 'messenger-particle') which issues from or passes through the origin can only reach points inside the cone of future or on its hyper-surface. That it cannot reach points which are definitely earlier (inside or on the cone of past) stands to reason. But also the region of virtual simultaneity is excluded, for in some frames they are earlier than the origin. There the message would arrive before it was sent out. That suffices to exclude the possibility in view of the principle of equal right to every Lorentz-frame. The necessary and sufficient condition is that no signal and no particle can ever move with a velocity greater than that of light.

I need hardly say, these consequences reached here rather dogmatically in our brief exposé formed in actual fact part of the basis on which the theory was built and they are amply sustained and corroborated by vast experimental evidence.

Considering the, to a certain degree, equal standing with the space coordinates that the time-coordinate is given already in the Special Theory of Relativity, the desire referred to in the beginning of this section crops up already in the Special Theory: to obtain a description of the velocity of a moving particle that is more in keeping with this attitude than are its three spatial components

$$\frac{dx_1}{dx_4}, \quad \frac{dx_2}{dx_4}, \quad \frac{dx_3}{dx_4}. \tag{10.6}$$

Now, if we call ds^2 the invariant (10.4) for two neighbouring world-points x_k and $x_k + dx_k$ of the path,

$$ds^2 = -dx_1^2 - dx_2^2 - dx_3^2 + dx_4^2, \tag{10.7}$$

and decide to give to ds (which is always real!) the sign of dx_4, then the four components

$$\frac{dx_1}{ds}, \quad \frac{dx_2}{ds}, \quad \frac{dx_3}{ds}, \quad \frac{dx_4}{ds} \tag{10.8}$$

are well qualified for the purpose. First, they transform in Lorentz-transformation exactly like the coordinates themselves, since this obviously holds for the dx_k, and ds is an invariant. They have an invariant of the same build, which trivially equals 1:

$$-\frac{dx_1^2}{ds^2} - \frac{dx_2^2}{ds^2} - \frac{dx_3^2}{ds^2} + \frac{dx_4^2}{ds^2} = 1. \tag{10.9}$$

Secondly, if we call the three components of the spatial velocity (10.6), for the moment, v_x, v_y, v_z, and v its absolute value, then

$$\frac{ds}{dx_4} = \sqrt{(1-v^2)} \qquad (10.10)$$

and the four components (10.8) read

$$\frac{v_x}{\sqrt{(1-v^2)}}, \quad \frac{v_y}{\sqrt{(1-v^2)}}, \quad \frac{v_z}{\sqrt{(1-v^2)}}, \quad \frac{1}{\sqrt{(1-v^2)}}. \qquad (10.11)$$

Hence in the very frequent case of the velocity being small compared with the velocity of light ($v \ll 1$), the first three components differ very little, only by second-order terms, from the components of the spatial velocity, and the fourth very little from unity.

In Special Relativity an array of four numbers which transforms *like the coordinates* is called a four-vector, and the four-vector (10.8) is referred to as the four-velocity. However the qualifying 'four-' is often dropped, if it can be understood from the context. From (10.11) it is seen that its components are not restricted to be < 1. Except for the condition (10.9), they can have any real values.

In making use of the notions of Special or Restricted Relativity for interpreting physically the mathematical scheme of General Relativity that we have been treating hitherto on these pages—treating it, to be sure, in a rather formal manner only—it cannot be emphasized too strongly that the latter is from a certain point of view not at all what its name seems to indicate; it is indeed from a certain point of view not a generalization but rather a restriction of the so-called Restricted Theory.

It restricts its validity to the infinitesimal neighbourhood of any—and that means, of course, of every—world-point. Practically the 'infinitesimal' region may often be taken fairly large, if the gravitational field in it is fairly weak and perceptibly constant, as, for example, within our laboratory for any length of time.

Nevertheless it is important to state that in principle we take over the scheme of Special Relativity only for a four-dimensional element of volume around a world-point, which may for that region play the part the 'origin' played in our foregoing deliberations. Coordinates, distances, time-intervals are the coordinate differentials or are formed of the coordinate differentials within that small region only.

A general transformation of the coordinates (x'_k = four arbitrary functions of the x_k) amounts to a linear transformation of the dx_k in each world-point, and thus amounts, as we shall see immediately, *inter alia* to a Lorentz-transformation of the dx_k at each point, varying continuously from point to point so that it can be regarded as constant within a small region. That is why we can take over the Special Theory for a small region.

Now in point of fact the transformation of the dx_k at any point

$$dx'_k = \frac{\partial x'_k}{\partial x_l} dx_l \qquad (10.12)$$

is *not* really a Lorentz-transformation, it is something slightly more general. A Lorentz-transformation, as regards the number of constants or of freedoms involved, is equivalent to a rotation in four dimensions. A rotation in n dimensions depends on $n(n-1)/1\cdot 2$ constants (giving 3 for $n = 3$, correctly as we know). For $n = 4$ we get 6 as the number of constants free in a Lorentz-transformation. (Another way of counting is: a Lorentz-transformation amounts to a rotation in space (3) plus an arbitrary velocity (3) of the new spatial frame with respect to the old one.)

But in (10.12) all the 16 coefficients $\partial x'_k/dx_l$ are arbitrary. What do the ten additional degrees of freedom amount to? Nothing very interesting from the point of view of this one world-point or world-element-of-volume. We would fain avoid them if we could. The general linear transformation—and that is what (10.12) is—includes four independent changes of the units in which the coordinates are measured in certain four mutually orthogonal directions;† it amounts to this plus an arbitrary 'rotation' (we defer the question whether or when the latter actually is a Lorentz-transformation).

That accounts for the ten additional degrees of freedom. Indeed we can choose the first of those four mutually orthogonal directions quite arbitrarily (3), then the second orthogonal to it (2), then the third orthogonal to both of them (1), then choose the four changes of measure (4), giving 10.

† Take here, for the moment, orthogonal in an elementary meaning. Discussion follows. For the counting of the constants it is immaterial.

We have to allow for these ten additional degrees of freedom in our world-point, simply because the general coordinate transformation seizes upon *all* world-points, and its differential form (10.12) could not possibly conform to a more special guise with only 6 constants at *every* world-point. And we do allow for this precisely by the fact that our invariant ds^2 has not the unchangeable form (10.7) in every frame

$$ds^2 = -dx_1^2 - dx_2^2 - dx_3^2 + dx_4^2, \tag{10.13}$$

but the more general form

$$ds^2 = g_{ik}\,dx_i\,dx_k, \tag{10.14}$$

the ten functions g_{ik} changing with the frame.

It cannot be denied that this is a great inconvenience. For from this general form we cannot even within this small region, *and not even with respect to the frame we are using*, tell the spatial distance and the time-interval between two point-events. *General world-coordinates are not fit for being interpreted directly*. If we wish to do that, we must reduce (10.14) by a local transformation to the form (10.13).

Let local coordinates for which ds^2 takes this standard form be called dy_k, the 'd' meaning only that they are infinitesimals, not necessarily that they are differentials of functions $y_k(x_l)$. For another system of world-coordinates x'_k, let dy'_k have the same meaning. Then the dy'_k are, via the dx'_k and dx_k, linear functions of the dy_k, and this linear connexion is bound to be a Lorentz-transformation at least in the sense that

$$-dy_1'^2 - dy_2'^2 - dy_3'^2 + dy_4'^2 = -dy_1^2 - dy_2^2 - dy_3^2 + dy_4^2.$$

Can we be sure that it is a Lorentz-transformation also in the restricted sense that its determinant $|\partial y'_k/\partial y_l|$ is $+1$ and that the coefficient $\partial y'_4/\partial y_4 > 0$?

The first can easily be guaranteed, if we agree once and for all that only world-transformations with positive determinant $|\partial x'_k/\partial x_i|$ shall be allowed—a small sacrifice, since the sign is constant anyhow—and if you use the same precaution for the transformation $dx_k \rightarrow dy_k$ that produces the standard form. The second, the positive sign of $\partial y'_4/\partial y_4$, *would be* guaranteed, if we could be sure that also in every general world-frame and at every world-point three of the four line-elements

$$\left.\begin{array}{lllll}(1) & (dx_1 & 0 & 0 & 0), \\ (2) & (0 & dx_2 & 0 & 0), \\ (3) & (0 & 0 & dx_3 & 0), \\ (4) & (0 & 0 & 0 & dx_4),\end{array}\right\} \qquad (10.15)$$

are space-like and one—moreover the same one throughout the world—time-like.

I am not aware that a world-frame has ever been used which does *not* comply with this demand. But I can see no general ground for, nor indeed a simple way of, excluding such a frame. It is not the case that the condition when complied with at one world-point must hold throughout the world. It is true that a coordinate-line-element† can change its character only by passing through the light-cone. But, unfortunately, that causes no singularity. The form

$$ds^2 = -dx_1^2 - dx_2^2 - 2dx_3\,dx_4$$

is just as regular and well-behaved as the standard form, into which it is readily turned by

$$dy_1 = dx_1 \quad dy_2 = dx_2 \quad dy_3 = \frac{1}{\sqrt{2}}(dx_3 + dx_4) \quad dy_4 = \frac{1}{\sqrt{2}}(dx_3 - dx_4).$$

Yet in the first form both $(0, 0, dx_3, 0)$ and $(0, 0, 0, dx_4)$ lie on the light-cone.

At some world-point near-by, dx_4 might be time-like, dx_3 space-like, at some other point the other way round. In the first we might feel inclined to interpret x_4 as the time, in the second x_3. Neither would be correct: the standard form must be produced for correct interpretation. This entails *inter alia* that the local velocity of light is everywhere the same in the whole world, viz. 1 in our units. But again we must not seriously regard this as a result of pure reason: the theory has been built *inter alia* on this demand.

We ought not to conclude this Chapter without indicating the point in which the General Theory actually *is* more general than the Restricted one. By a Lorentz-transformation we can 'transform away' any velocity. Can the general theory do more, since for interpretation we eventually have to fall back on a frame of restricted

† We mean one of the type (10.15).

relativity, from which the local g_{ik} have, so to speak, disappeared? Are not they supposed to depict the gravitational field?

Not they but, as we shall soon see, their first derivatives. And we shall also see that they too can be made zero at any given point by a suitable *general* transformation. We thus 'transform away' the local gravitational field. Not by a conjuring trick: the physical meaning is that we adopt locally a spatial coordinate system which shares the *acceleration* a test-body would experience in the local gravitational field.

Note added in proof. In explaining on p. 85 that there is no necessity for just three of the four line-elements (10.15) being space-like, one time-like, I added that I was not aware of anybody ever using a frame in which this was not so. Since this was written, an example to the contrary was furnished by Kurt Gödel, *Reviews of Modern Physics*, **21**, 447, 1949. Gödel communicates a fascinating, entirely novel type of cosmological solution of Einstein's 1915 theory. This solution acquires its simplest form with *two* of the coordinate-line-elements time-like (the other two space-like). As far as I can see, it is in this case not possible to find a frame such that the customary distribution (3+1) holds *everywhere*. The *signature* of Gödel's line-element is, of course, $(---+)$, as required!

CHAPTER XI

CONSERVATION LAWS AND VARIATIONAL PRINCIPLES

THE ELEMENTARY NOTION OF CONSERVATION LAWS

We have proposed as the field-laws of gravitation

$$\left.\begin{aligned} R_{kl} &= 0 \quad \text{in empty space-time,} \\ R_{kl} &= T_{kl} \quad \text{where there is `matter',} \end{aligned}\right\} \qquad (11.1)$$

T_{kl} being the stress-energy-momentum tensor of matter. (What it means exactly will be discussed forthwith.)

One must be very careful in adopting field equations at pleasure, just as much as, or even more than, when writing down a set of algebraic equations to determine some unknown quantities. For they need not be compatible. For instance, if you demand of x, y

$$x+y = 1,$$
$$2x+2y = 5,$$

this is just not possible.

A second, equally important, but quite different, point is that matter has to fulfil the four conservation laws of energy and linear momentum. That means, as I shall explain forthwith, a certain condition which T_{kl} has to fulfil, namely that its divergence must vanish. We must not demand it to equal another tensor for which this is not the case. (The second of the equations (11.1) will actually have to be modified on this account.)

Both requirements—the compatibility and the conservation laws—are automatically fulfilled if we do not accept our field equations straight away, but base them on a variational principle. This is a programme the consummation of which will ocupy us for some time. I do not mean the question of compatibility; we shall take it for a granted mathematical theorem that the so-called Euler-equations, the variational equations deriving from a variational principle, are always compatible. But we have to go to some length about the conservation laws. I shall first speak of elementary cases, quite apart from General Relativity.

The prototype of a conservation law is the so-called equation of continuity in the motion of a fluid. If ρ is the density and v_1, v_2, v_3 the components of the velocity, this equation reads

$$\frac{\partial \rho}{\partial t}+\frac{\partial \rho v_1}{\partial x_1}+\frac{\partial \rho v_2}{\partial x_2}+\frac{\partial \rho v_3}{\partial x_3}=0. \tag{11.2}$$

By integrating it over some volume of the fluid, a volume fixed in space, and using Gauss's theorem you get

$$-\frac{d}{dt}\int \rho d\tau = \int [\rho v_1 \cos(n, 1)+\rho v_2 \cos(n, 2)+\rho v_3 \cos(n, 3)]\, df.$$

The three-dimensional vector ρv is the density of flux; the equation states that the amount of fluid contained in that space diminishes by the amount that flows out. It is thus a very trivial but none-the-less valuable statement of a purely geometrical nature; it has nothing to do with the dynamical interaction between the parts of the fluid or what not.

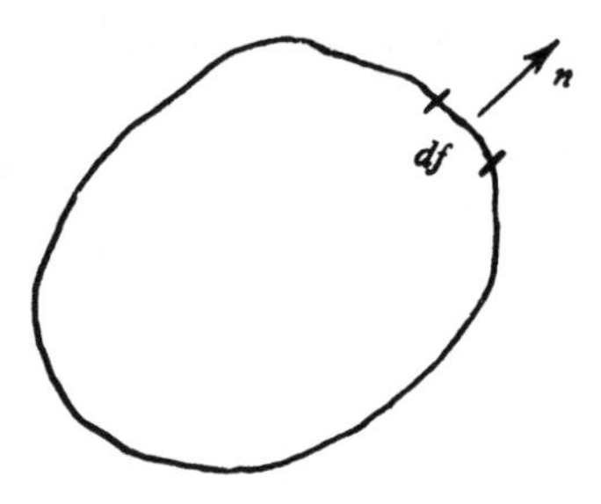

In several elementary theories, to wit in the motion of an ideal fluid or of an ideal elastic body or in Maxwell's theory, a similar consideration applies to the dynamical quantities of density of energy and density of momentum. Each of the *four* quantities, density of energy and x-, y-, z-component of the density of momentum, can take the rôle of our ρ above, and so we have *four* dynamical equations of conservation of similar build to equation (11.2). With each of the four quantities is associated a triplet of components indicating the *flux* of that quantity, so that we have 16 quantities all in all. The three components giving the flux of a component of the 3-vector of momentum can, of course, not form a 3-vector. The 3×3 components of momentum flux form a single entity, the three-dimensional stress-tensor. Let us call M_k $(k = 1, 2, 3)$ the components of momentum and T_{ik} $(i, k = 1, 2, 3)$ this stress-tensor, then the conservation equation for M_1 reads

$$-\frac{\partial M_1}{\partial t}+\frac{\partial T_{11}}{\partial x_1}+\frac{\partial T_{12}}{\partial x_2}+\frac{\partial T_{13}}{\partial x_3}=0,$$

which gives by integration over a volume

$$\frac{d}{dt}\int M_1\,d\tau = \int [T_{11}\cos(n,1) + T_{12}\cos(n,2) + T_{13}\cos(n,3)]\,df.$$

The second integrand is the x-component of the force, F_1, exerted from the immediate neighbourhood outside on the immediate neighbourhood inside df. The equation says that this force contributes to—and the totality of these forces make up—the increase of the total x-component of momentum inside the surface. That need not necessarily result in motion, because it is possible that all these stresses balance and that there is equilibrium. But one regards this force in any case as a flux of momentum. Moreover, there may be an intrinsic hidden motion going on and producing a convective transfer of momentum across the element of surface. It is sometimes convenient to include this in the stress. For instance, in the interior of a gas we always do that. There are all the time molecules crossing over in the direction 1 → 2 and carrying, on the whole, momentum *of* this direction *in* this direction. *And this is by no means counter-balanced by the opposite events*; on the contrary, it is doubled. For the particles crossing over from (2) to (1) carry, on the whole, momentum of the *opposite* sign in the *opposite* direction. We are wont to refer to the phenomenon as the internal pressure of the gas.

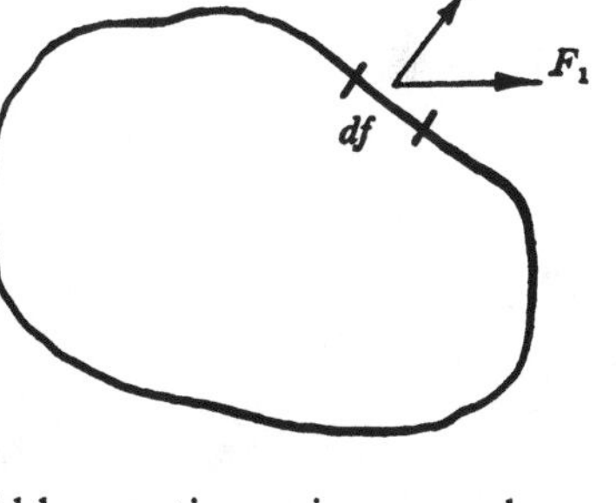

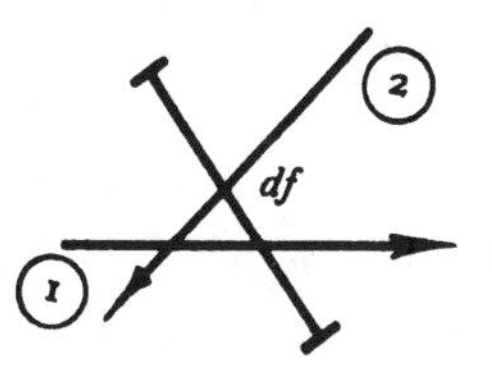

We must now take a pretty wide leap, to avoid which would lead us to great length, for it would really mean building up Maxwell's theory of the electromagnetic field from its experimental foundations. So I must ask the reader to believe me that there are strong grounds from our four-dimensional standpoint for comprising the four conservation equations into *one* statement about one four-dimensional tensor of the second rank. Its 16 components are, broadly speaking, the 16 quantities mentioned just before, but they are not really 16, only 10, because the tensor is symmetric.

(With the 9 stress-components it is a well-known fact, turning up for the first time in the theory of elasticity, that they form a symmetrical tensor and thus amount only to 6.) We have to use the so-called Galilean metric of Special Relativity ($g_{ik} = -1, -1, -1, +1$ on the diagonal, otherwise zero). Then the most uniform shape is obtained by using the mixed tensor, because that avoids minus signs. We then have

$$\left.\begin{aligned}
\frac{\partial T_1^1}{\partial x_1}+\frac{\partial T_1^2}{\partial x_2}+\frac{\partial T_1^3}{\partial x_3}+\frac{\partial T_1^4}{\partial x_4} &= 0,\\
\frac{\partial T_2^1}{\partial x_1}+\frac{\partial T_2^2}{\partial x_2}+\frac{\partial T_2^3}{\partial x_3}+\frac{\partial T_2^4}{\partial x_4} &= 0,\\
\frac{\partial T_3^1}{\partial x_1}+\frac{\partial T_3^2}{\partial x_2}+\frac{\partial T_3^3}{\partial x_3}+\frac{\partial T_3^4}{\partial x_4} &= 0,\\
\frac{\partial T_4^1}{\partial x_1}+\frac{\partial T_4^2}{\partial x_2}+\frac{\partial T_4^3}{\partial x_3}+\frac{\partial T_4^4}{\partial x_4} &= 0.
\end{aligned}\right\} \qquad (11.3)\dagger$$

The most spectacular new event is that the three components of momentum are at the same time those of the flux of energy. This became clear for the first time in Maxwell's theory, where these three components form the so-called Poynting-three-vector. The deeper significance is that energy and mass is the same thing. Momentum, in its original conception mass × velocity, is a stream of mass, and thus of energy. So the last of our four equations is really at the same time the equation of continuity of mass-density and brings us back to the point from which we started.

What form those equations will have to take in a general metric, that is, in a general gravitational field, must be left open for the moment. We note that we shall have to have something like the vanishing of the divergence of a tensor (or probably tensor density) of the second rank. *This* we shall have to interpret as the conservation laws in a gravitational field.

† Or briefly $\dfrac{\partial T^i{}_k}{\partial x_i} = 0.$

HOW CONSERVATION LAWS FOLLOW FROM A VARIATIONAL PRINCIPLE IN CLASSICAL (PRE-RELATIVISTIC) THEORIES

We shall now make a mathematical study of the way in which a set of equations of this kind emerge from a variational principle. There is one old and well-known way—you find it in every text-book of variational calculus.

We begin by sketching this elementary method, not so much for the purpose of *using* it later as, inversely, for *contrasting* it with the method used in General Relativity; this is intrinsically different, yet bears some outward resemblance to the elementary one, so that the reader might easily confuse it with the latter. The elementary connexion runs as follows.

Let
$$f, g, h \ldots$$
be a set of undetermined functions of the coordinates x_k (in the most relevant application it will be the field-variables g_{ik} or something like that). Write for abbreviation

$$\frac{\partial f}{\partial x_k} = f_k, \text{ etc.}$$

We envisage the variation δI of the four-dimensional integral

$$I = \int H(f, g, h, \ldots; f_k, g_k, h_k, \ldots; x_k)\, dx^4,$$

taken over any fixed region. H is to be a *given* function of its many arguments ($5s+4$, if there are s functions $f, g, h, \ldots$). By variation we mean that $f, g, h, \ldots$ are given small increments $\delta f, \delta g, \delta h, \ldots$, which *vanish* at the boundary.

For convenience, partial derivatives of H as a function of its $5s+4$ arguments will be indicated by subscripts thus:

$$\frac{\partial H}{\partial f} = H_f \ldots \frac{\partial H}{\partial f_k} = H_{f_k} \ldots \frac{\partial H}{\partial x_k} = H_k.$$

The summation sign Σ, preceding an expression in which the letter f occurs, will mean: plus the same expression for g, for h, etc. Our *usual* summation rule is retained! Then†

$$\delta I = \int \Sigma(H_f \delta f + H_{f_k} \delta f_k)\, dx^4 = \int \Sigma \left(H_f - \frac{\partial}{\partial x_k} H_{f_k} \right) \delta f\, dx^4.$$

† We use that $\delta f_k = \partial/\partial x_k\, (\delta f)$, perform a partial integration in the term containing this quantity, and remember $\delta f = 0$ at the boundary, so that the surface integral vanishes.

The bracket is sometimes called the *Hamiltonian derivative* of H with respect to f, and sometimes abbreviated $\delta H/\delta f$ or similarly. For δI to vanish under our conditions of variation *all* these Hamiltonian derivatives must vanish. The system of these equations is called the Euler equations:

$$\frac{\delta H}{\delta f} \equiv H_f - \frac{\partial}{\partial x_k} H_{fk} = 0, \quad \Big| \quad f_i$$

$$\frac{\delta H}{\delta g} \equiv H_g - \frac{\partial}{\partial x_k} H_{gk} = 0, \quad \Big| \quad g_i$$

$$\cdots\cdots\cdots\cdots\cdots\cdots \quad \Big| \quad \cdots$$

Now if you multiply the first equation by f_i, the second by g_i, etc. and add them up, you get (remember the meaning of Σ):

$$0 = \Sigma\left(f_i H_f - f_i \frac{\partial}{\partial x_k} H_{fk}\right) = \Sigma\left[f_i H_f + \frac{\partial f_i}{\partial x_k} H_{fk} - \frac{\partial}{\partial x_k}(f_i H_{fk})\right].$$

Now since $f_i = \dfrac{\partial f}{\partial x_i}$, $\dfrac{\partial f_i}{\partial x_k} = \dfrac{\partial f_k}{\partial x_i}$, the Σ over the first two terms will give simply $\partial H/\partial x_i$ provided H does not contain x_i explicitly. (We have assumed this explicit dependence only in order to *drop* it now: to show that the whole argument turns on this non-dependence!) So our equation can be written

$$0 = \frac{\partial H}{\partial x_i} - \frac{\partial}{\partial x_k} \Sigma(f_i H_{fk}).$$

Using Kronnecker's symbol δ_{ik} ($= 1$ for $i = k$, otherwise 0) we may write

$$\frac{\partial}{\partial x_k}(\delta_{ik} H - \Sigma f_i H_{fk}) = 0.$$

These are four equations of the 'conservation-type'. Note three things:

(i) They are *consequences* of the Euler-equations; they hold if and only if $\delta I = 0$.

(ii) They require *severally* that H shall not explicitly depend on the x_i in question.

(iii) The brackets are not necessarily symmetric in i, k. In general they are *not*.

In non-relativistic or pre-relativistic theories this was the usual way of accounting for the conservation laws. That they hinge on H

not depending directly on the x_k was very significant. The Hamiltonian function (often called Lagrange-function) H dictates the laws of our field. The independence on x_k means that these laws are not affected by a displacement in space or time, that they are the same on Monday as on Tuesday, the same in Paris as in London. This independence is (according to pre-relativistic physics) at the basis of the conservation of energy and momentum.

CONSERVATION LAWS IN GENERAL RELATIVITY

In General Relativity things are changed. General Relativity itself entails conservation laws—and that not as consequences of field equations, but as *identities*. What happens is simply this: If you contemplate an integral

$$I = \int \mathfrak{R}\, dx^4$$

in which $\mathfrak{R}$ is now definitely assumed to be an invariant density (and therefore written in Gothic), then from the mere fact of the general invariance of this integral follow four *identical* relations between the *Hamiltonian derivatives* of $\mathfrak{R}$, relations of the type of conservation laws; identities, as I said; not, as before, equations which result from putting the Hamiltonian derivatives equal to zero; four relations between the Hamiltonian derivatives will be shown to hold whether or not these derivatives themselves are zero; indeed if they are, the relations become *trivial*.

The existence of four identities is not astonishing. For the following reason. *If you did adopt* the variational principle $\delta I = 0$ to determine the functions $f, g, h, \ldots$ (on which and on whose derivatives $\mathfrak{R}$ shall depend), you would get the differential equations

$$\frac{\delta \mathfrak{R}}{\delta f} = 0, \text{ etc.}$$

as many as there are functions $f, g, h, \ldots$. And that is, in General Relativity, just by four too many; because here the functions $f, g, h, \ldots$ are bound to be tensor-components and are thus amenable to change under a general transformation of the frame. Such a transformation contains four arbitrary functions. Hence our s functions cannot be fully controlled by those s equations.

Hence the latter cannot be quite independent of each other. The dependence is constituted precisely by those four identities between their first members.

We proceed to derive them in the case of a metric g_{ik}. That is to say we take the scalar density $\mathfrak{R}$ now to depend only on the ten g_{ik} (which now take the part of $f, g, h, \ldots$), their derivatives $g_{ik,l}$ and possibly also on higher derivatives. (This involves no complication and will be required later.) Thus we contemplate

$$\mathfrak{R}(g_{ik}, g_{ik,l}, g_{ik,l,m}, \ldots).$$

We compute δI and perform, after the pattern indicated before, the partial integrations needed to put it into the form

$$\delta I = \int \frac{\delta \mathfrak{R}}{\delta g_{ik}} \delta g_{ik} \, dx^4. \tag{11.4}$$

For our present purpose—which is *not* to demand $\delta I = 0$ but to derive identities between the $\delta\mathfrak{R}/\delta g_{ik}$—it is not necessary actually to compute the latter, it suffices to know that they can easily be determined. For shortness we put

$$\frac{\delta \mathfrak{R}}{\delta g_{ik}} = \mathfrak{R}^{ik},$$

thus

$$\delta I = \int \mathfrak{R}^{ik} \delta g_{ik} \, dx^4. \tag{11.5}$$

δI is an invariant, being the difference of two invariants. Hence the integrand is a scalar density. Moreover, δg_{ik} is a tensor, being the difference of two tensors, and is entirely arbitrary. Hence $\mathfrak{R}^{ik}$ is a symmetrical contravariant density of the second rank.† Our notation has been chosen accordingly.

We now apply (11.5) to a variation δg_{ik} for which δI must be identically zero, on the strength of I being an invariant. To wit, we bring about a 'variation' of the g_{ik} merely by a change of frame (which cannot alter the value of the invariant I). We let

† There would be no point in distinguishing in (11.4) between $\delta\mathfrak{R}/\delta g_{ik}$ and $\delta\mathfrak{R}/\delta g_{ki}$, because they coalesce in view of the symmetry of δg_{ik}. Hence $\mathfrak{R}^{ik}$ is of necessity symmetric, and that is why the arbitrariness of the *symmetrical* factor δg_{ik} suffices to prove the tensor-property of $\mathfrak{R}^{ik}$. Our general consideration on p. 12 was based on an arbitrary co-factor of the form $A_i A_k$, with A_i arbitrary. For an arbitrary *tensor* A_{ik} (as δg_{ik} is) they hold *a fortiori*.

the transformation depend on a parameter λ so that for $\lambda \to 0$ it approaches to identity:

$$x_l = x_l(x'_i, \lambda) = x'_l + \lambda\phi_l(x'_i) + \lambda^2\psi_l(x'_i) + \dots.$$

Our invariant integral then takes the form

$$I = \int \mathfrak{R}\left(\frac{\partial x_l}{\partial x'_i}\frac{\partial x_m}{\partial x'_k} g_{lm}(x_r), \dots\right) dx'^4,$$

where $\mathfrak{R}$ is, of course, *the same function* of the

$$g'_{ik}(x'_s) = \frac{\partial x_l}{\partial x'_i}\frac{\partial x_m}{\partial x'_k} g_{lm}(x_r),$$

and of their derivatives with respect to the x', as it was before of the g_{ik} and their derivatives w.r.t. the x. The limits of integration in x' will also be *the same* as they were in x, if we now stipulate that at the boundary the transformation shall approach to identity for any λ. Since the *notation* for the integration variable is irrelevant, the only *formal* change is that the *argument* is now not $g_{ik}(x'_s)$ but $g'_{ik}(x'_s)$. Now, you easily compute, by expanding with respect to λ, that

$$g'_{ik}(x'_s) - g_{ik}(x'_s) = \lambda\left(g_{lk}\frac{\partial\phi_l}{\partial x'_i} + g_{im}\frac{\partial\phi_m}{\partial x'_k} + \frac{\partial g_{ik}}{\partial x'_n}\phi_n\right) + O(\lambda^2),$$

all functions to be written with the x'. Since this must vanish at the border for any λ, we may use (11.5) with the first order terms in λ for δg_{ik}. Dropping the now superfluous dashes we get

$$\delta I = \int \mathfrak{R}^{ik}\left(g_{lk}\frac{\partial\phi_l}{\partial x_i} + g_{im}\frac{\partial\phi_m}{\partial x_k} + \frac{\partial g_{ik}}{\partial x_n}\phi_n\right) dx^4 = 0.$$

We use the abbreviation of index-pulling by the g_{ik} and perform the suggested partial integrations

$$0 = \int\left(-\frac{\partial\mathfrak{R}^i{}_l}{\partial x_i}\phi_l - \frac{\partial\mathfrak{R}^k{}_m}{\partial x_k}\phi_m + \mathfrak{R}^{ik}\frac{\partial g_{ik}}{\partial x_n}\phi_n\right) dx^4$$

$$= \int\left(-2\frac{\partial\mathfrak{R}^k{}_m}{\partial x_k} + \mathfrak{R}^{ik}\frac{\partial g_{ik}}{\partial x_m}\right)\phi_m\, dx^4.$$

Now ϕ_m is quite arbitrary, hence

$$\frac{\partial\mathfrak{R}^k{}_m}{\partial x_k} - \tfrac{1}{2}\mathfrak{R}^{ik}\frac{\partial g_{ik}}{\partial x_m} = 0.$$

These are the four identities. They must, of course, be tensor-equations. We can shew this directly by turning them into our semicolon notation. Indeed (the two terms we add are skew in (i, k) and thus vanish on summation over the symmetrical $\mathfrak{K}^{ik}$):

$$\tfrac{1}{2}\mathfrak{K}^{ik}\frac{\partial g_{ik}}{\partial x_m} = \tfrac{1}{2}\mathfrak{K}^{ik}\left(\frac{\partial g_{im}}{\partial x_k}+\frac{\partial g_{ik}}{\partial x_m}-\frac{\partial g_{km}}{\partial x_i}\right)$$

$$= \mathfrak{K}^k{}_i\begin{Bmatrix} i \\ k\ m \end{Bmatrix}.$$

Thus we get $$\frac{\partial \mathfrak{K}^k{}_m}{\partial x_k} - \mathfrak{K}^k{}_i\,\Gamma^i{}_{km} = \mathfrak{K}^k{}_{m;k} = 0.$$

(From the general formula for $\mathfrak{K}^k{}_{m;l}$ which has *three* supplementary Γ-terms; but two of them cancel on contraction (k, l).)

Of course, from the rule of pulling under the semicolon you may also write

$$\mathfrak{K}^{km}{}_{;k} = 0.$$

So the simple fact is: *the invariant divergence vanishes for the tensor density that is constituted by the Hamiltonian derivatives of any scalar density* $\mathfrak{K}$ *that depends only on the* g_{ik} *and their derivatives with respect to the coordinates up to any finite order.*

That is, of course, very nice, because the invariant derivative is doubtless the counterpart, the only invariant counterpart, of the ordinary divergence in elementary theory. And so we have good hope that some suitable scalar density will yield us the conservation laws as identities in the general theory. Yet a few remarks must be made to damp our enthusiasm. The fairy gift we are presented with contains a little too much and a little too little.

First, these identities are not unique. *Every* scalar density produces a set. We might perhaps have preferred to get a statement about one particular $\mathfrak{K}$ only, even if it were not an identity. We are not really so keen on getting identities, while on the other hand the conservation laws are one individual fact, not a class of facts.

Secondly, there *appears* to be the following alternative. Either the $\mathfrak{K}$ from which the 'true' conservation laws arise is the same as the one that will yield our field equations, or it is not the same. In the first case, the conservation laws become trivial, because the $\mathfrak{K}^{km}$ vanish separately, not only their divergence; in the second case, when there are two different $\mathfrak{K}$'s for the two purposes, it would seem that the conservation laws have nothing to do with the field equations at all, they would stand quite aloof.

A third somewhat disparaging remark is this. Though the invariant divergence is the only possible truly invariant counterpart of the elementary one, it is not a true divergence in the mathematical sense. You cannot by integrating over a three-dimensional volume derive the integrated conservation laws, which are really those to appeal immediately to imagination.

A merely formal remark, that is quite useful, is this. We had put

$$\delta I = \int \mathfrak{R}^{ik}\,\delta g_{ik}\,dx^4.$$

What if we had preferred to use the functions g^{ik} rather than the g_{ik}? That is simple enough. Both factors are tensors and so we can raise and lower any pair of dummies simultaneously. Moreover, we know that by raising both indices in δg_{ik} we get $-\delta g^{ik}$. Hence

$$\delta I = -\int \mathfrak{R}_{ik}\,\delta g^{ik}\,dx^4.$$

In other words

$$\frac{\delta \mathfrak{R}}{\delta g^{ik}} = -\mathfrak{R}_{ik}.$$

EINSTEIN'S VARIATIONAL PRINCIPLE

The simplest scalar density you can make up of the g_{ik}'s is $\sqrt{-g}$. So let us just look for a moment at

$$I = \int \sqrt{-g}\,dx^4. \tag{11.6}$$

Now

$$\frac{\delta g}{g} = g^{ik}\,\delta g_{ik} = \frac{\delta(-g)}{-g} = \delta \lg(-g) = 2\delta \lg\sqrt{-g} = 2\frac{\delta\sqrt{-g}}{\sqrt{-g}}.$$

So

$$\delta I = \tfrac{1}{2}\int \sqrt{-g}\,g^{ik}\,\delta g_{ik}\,dx^4.$$

Therefore the Hamiltonian derivative

$$\frac{\delta\sqrt{-g}}{\delta g_{ik}} = \tfrac{1}{2}\sqrt{-g}\,g^{ik} = \tfrac{1}{2}\mathfrak{g}^{ik}.$$

But $\mathfrak{g}^{ik} = 0$ cannot serve as a field equation. Moreover the identities $\mathfrak{g}^{ik}{}_{;k} = 0$ are trivial understatements. For we know that even $\mathfrak{g}^{ik}{}_{;l} = 0$.

The next complicated scalar density is already very much more complicated, it is formed of the curvature scalar

$$R = g^{ik} R_{ik}$$

by multiplying it by $\sqrt{-g}$. Thus we now contemplate

$$I = \int R\sqrt{-g}\, dx^4 = \int \mathfrak{g}^{ik} R_{ik}\, dx^4. \tag{11.7}$$

It is indeed much more complicated, because the R_{ik} depend on second derivatives of the g_{ik}. You would expect the Hamiltonian derivatives to reach the fourth order. In actual fact the variation is very simple to perform, and the Hamiltonian derivatives are only of the second order. We first get

$$\delta I = \int (\delta \mathfrak{g}^{ik} R_{ik} + \mathfrak{g}^{ik} \delta R_{ik})\, dx^4.$$

Now remember the Palatini equation (6.19)

$$\delta R_{ik} = -(\delta \Gamma^{\alpha}{}_{ik})_{;\alpha} + (\delta \Gamma^{\alpha}{}_{i\alpha})_{;k}.$$

If you insert this and perform the partial integrations *with respect to the semicolons* the terms with δR_{ik} give nothing, because $\mathfrak{g}^{ik}{}_{;\alpha} \equiv 0$. So we are left with

$$\delta I = \int R_{ik} \delta \mathfrak{g}^{ik}\, dx^4. \tag{11.8}$$

Since we may, of course, regard the $\mathfrak{g}^{ik}$ as independent variables (just as well as the g_{ik} or g^{ik}) the Euler equations are $R_{ik} = 0$. These are just our proposed field equations. What about the four identities? Well, *they* hold for the Hamiltonian derivatives with respect to the g_{ik} or g^{ik}. We must therefore express δI by them. That is easily done:

$$\delta \mathfrak{g}^{ik} = \delta \sqrt{-g}\, g^{ik} = \sqrt{-g}\, \delta g^{ik} + g^{ik} \delta \sqrt{-g},$$

$$\delta \sqrt{-g} = \tfrac{1}{2} \sqrt{-g}\, g^{ik} \delta g_{ik} = -\tfrac{1}{2} \sqrt{-g}\, g_{ik} \delta g^{ik},$$

$$\delta \mathfrak{g}^{ik} = \sqrt{-g}\, \delta g^{ik} - \tfrac{1}{2} g^{ik} \sqrt{-g}\, g_{\mu\nu} \delta g^{\mu\nu}.$$

Hence

$$\delta I = \int R_{ik} \sqrt{-g}\, (\delta g^{ik} - \tfrac{1}{2} g^{ik} g_{\mu\nu} \delta g^{\mu\nu})\, dx^4$$

$$= \int \sqrt{-g}\, (R_{ik} \delta g^{ik} - \tfrac{1}{2} R g_{ik} \delta g^{ik})\, dx^4$$

$$= \int \sqrt{-g}\, (R_{ik} - \tfrac{1}{2} g_{ik} R)\, \delta g^{ik}\, dx^4.$$

So
$$\frac{\delta R\sqrt{-g}}{\delta g^{ik}} = \sqrt{-g}\,(R_{ik} - \tfrac{1}{2}g_{ik}R). \tag{11.9}$$

And the identities read

$$\left.\begin{aligned} &[\sqrt{-g}\,(R^i{}_k - \tfrac{1}{2}\delta^i{}_k R)]_{;i} = 0, \\ \text{or}\quad &(R^i{}_k - \tfrac{1}{2}\delta^i{}_k R)_{;i} = 0, \\ \text{or}\quad &(R^{ik} - \tfrac{1}{2}g^{ik}R)_{;i} = 0, \end{aligned}\right\} \tag{11.10}$$

all that meaning the same thing.

This shows us that we must certainly not regard R_{ik} as the matter tensor, but $R_{ik} - \frac{1}{2}g_{ik}R$. There will, of course, be some constant factor, essentially the constant of gravitation. We leave it out at the moment, except that for reasons I cannot explain at the moment the factor is negative. Thus in places where there is matter we shall have to put

$$-(R_{ik} - \tfrac{1}{2}g_{ik}R) = T_{ik}. \tag{11.11}$$

I would rather you did not regard these equations as field equations, but as a definition of T_{ik}, the matter tensor. Just in the same way as Laplace's equation $\operatorname{div} E = \rho$ (or $\nabla^2 V = -4\pi\rho$) says nothing but: wherever the divergence of E is *not* zero we say there is a charge and call $\operatorname{div} E$ the density of charge. Charge does not *cause* the electric vector to have a non-vanishing divergence, it *is* this non-vanishing divergence. In the same way, matter does not *cause* the geometrical quantity which forms the first member of the above equation to be different from zero, it *is* this non-vanishing tensor, it is described *by it*.

NON-INVARIANT FORM OF THE CONSERVATION LAWS

We make an explicit note of the following Hamiltonian derivatives:

$$\frac{\delta R\sqrt{-g}}{\delta \mathfrak{g}^{ik}} = R_{ik}, \tag{11.12}$$

$$\frac{\delta R\sqrt{-g}}{\delta g^{ik}} = \sqrt{-g}\,(R_{ik} - \tfrac{1}{2}g_{ik}R) = -\sqrt{-g}\,T_{ik} = -\mathfrak{T}_{ik}, \tag{11.13}$$

$$\frac{\delta R\sqrt{-g}}{\delta g_{ik}} = \mathfrak{T}^{ik}. \tag{11.14}$$

The first is drawn from (11.7) and (11.8), the second from (11.9) with the notation (11.11), the third from δg^{ik} being the associate of $-\delta g_{ik}$ (see p. 72). $\mathfrak{T}_{ik}$ is the tensor T_{ik} 'gothicized' in the ordinary way. The first form of the identities (11.10) can then be written

$$\mathfrak{T}^i{}_{k;\,i} = 0. \tag{11.15}$$

We shall also use it in the more explicit form

$$\frac{\partial \mathfrak{T}^i{}_k}{\partial x_i} - \tfrac{1}{2}\mathfrak{T}^{lm}\frac{\partial g_{lm}}{\partial x_k} = 0, \tag{11.16}$$

actually the first we had obtained in subsect. 3, where we dealt with an arbitrary invariant density $\mathfrak{R}$ (see the equation at the bottom of p. 95).

The conservation identities (11.15), while extremely satisfactory from the point of view of general invariance, lack the simple visualizable meaning we had explained in the elementary cases: three-dimensional volume integrals of their first members cannot immediately by partial integration be transformed in such a way as to allow us to interpret, for example, $\mathfrak{T}^1_1$, $\mathfrak{T}^1_2$, $\mathfrak{T}^1_3$ as the flux of the quantity of which $\mathfrak{T}^1_4$ is the density, etc. The second term in (11.16) is in the way. One might object—or rather acquiesce by the excuse—that the three-dimensional volume integral of the component of a vector density or tensor density has no invariant meaning anyhow. Still we should welcome a visualizable interpretation conforming to the elementary way of thinking, albeit only in a particular frame and perhaps for a not too extended region. Our main objective in the following is to remove the obstacle by turning also the second term in (11.16) into a sum of derivatives with respect to the coordinates—what is sometimes called a 'plain divergence'.

We begin by exhibiting the reason why the Hamiltonian derivatives (H.D., for brevity) collected in (11.12)–(11.14) do not depend on higher than second derivatives of the fundamental tensor, while from elementary variational calculus we should expect them to reach the fourth order, viz. twice the order they reach in the integrand $R\sqrt{-g}$ itself. The reason is that the latter is in a certain sense *equivalent* to another integrand, to be called $-\mathfrak{L}$ in the following, which contains no higher than first derivatives of the g_{ik}. The equivalence rests on this, that the *difference* $R\sqrt{-g}+\mathfrak{L}$ is what we just before called a 'plain divergence'. The (four-dimensional) integral of a plain divergence, since it can be turned into a (hyper-) surface integral, suffers no change under variations such as served us to define the H.D., viz. variations that vanish at the boundary. Hence the H.D. of a plain divergence vanishes. Applying this to our case we see that any H.D. of $R\sqrt{-g}$ is equal to the corresponding one of $\mathfrak{L}$, taken with the negative sign. This is what we really mean by calling the integrands $R\sqrt{-g}$ and $-\mathfrak{L}$ equivalent. We proceed to fill in the details.

Envisage the explicit expression

$$R\sqrt{-g} = \mathfrak{g}^{ik}R_{ik} = \mathfrak{g}^{ik}\left(-\frac{\partial \Gamma^{\alpha}_{ik}}{\partial x_{\alpha}}+\frac{\partial \Gamma^{\alpha}_{i\alpha}}{\partial x_{k}}+\Gamma^{\beta}_{\alpha k}\Gamma^{\alpha}_{i\beta}-\Gamma^{\beta}_{\alpha\beta}\Gamma^{\alpha}_{ik}\right).$$

Let me introduce the abbreviations

$$\Gamma^{\beta}_{\alpha k}\Gamma^{\alpha}_{i\beta}-\Gamma^{\beta}_{\alpha\beta}\Gamma^{\alpha}_{ik} = \mathfrak{L}_{ik}, \qquad (11.17)$$

$$\mathfrak{g}^{ik}\mathfrak{L}_{ik} = \mathfrak{L}.$$

Thus $$R\sqrt{-g} = -\mathfrak{g}^{ik}\frac{\partial \Gamma^{\alpha}_{ik}}{\partial x_{\alpha}}+\mathfrak{g}^{ik}\frac{\partial \Gamma^{\alpha}_{i\alpha}}{\partial x_{k}}+\mathfrak{L}.$$

By adding to this a certain plain divergence you get the equivalent integrand

$$\mathfrak{g}^{ik}{}_{,\alpha}\Gamma^{\alpha}_{ik}-\mathfrak{g}^{ik}{}_{,k}\Gamma^{\alpha}_{i\alpha}+\mathfrak{L}.$$

Express the factors $\mathfrak{g}^{ik}{}_{,\alpha}$ and $\mathfrak{g}^{ik}{}_{,k}$ by those linear aggregates of Γ's to which they are equal in virtue of the fact that $\mathfrak{g}^{ik}{}_{;\alpha}$ vanishes identically. Of the six terms you thus obtain two cancel, the remaining four give $-2\mathfrak{L}$:

$$\mathfrak{g}^{ik}{}_{,\alpha}\Gamma^{\alpha}_{ik}-\mathfrak{g}^{ik}{}_{,k}\Gamma^{\alpha}_{i\alpha} = -2\mathfrak{L}. \qquad (11.18)$$

So the equivalent integrand reduces to $-\mathfrak{L}$, as announced.

It is *not* an invariant density. Its advantage is precisely that it contains no higher than first derivatives. You may regard it as a function of either the g_{ik} or the g^{ik} or the $\mathfrak{g}^{ik}$ or the $\mathfrak{g}_{ik}$ and of their coordinate-derivatives *respectively*, that is of either $g_{ik,l}$ or $g^{ik}{}_{,l}$ or $\mathfrak{g}^{ik}{}_{,l}$ or $\mathfrak{g}_{ik,l}$ according to your first choice. In each case its H.D.'s can be written down straight away along the lines explained in the subsect., pp. 91 f. They are alternative expressions, whose merits will appear forthwith, of the H.D.'s of $R\sqrt{-g}$ indicated above, to which they are respectively equal.

We have for instance

$$\mathfrak{T}^{lm} = -\frac{\delta\mathfrak{L}}{\delta g_{lm}} = \frac{\partial}{\partial x_{\alpha}}\left(\frac{\partial\mathfrak{L}}{\partial g_{lm,\alpha}}\right)-\frac{\partial\mathfrak{L}}{\partial g_{lm}}.$$

Let us use this in the second term of (11.16)

$$\mathfrak{T}^{lm}\frac{\partial g_{lm}}{\partial x_k} = -\frac{\delta\mathfrak{L}}{\delta g_{lm}}\frac{\partial g_{lm}}{\partial x_k} = \frac{\partial}{\partial x_\alpha}\left(\frac{\partial\mathfrak{L}}{\partial g_{lm,\alpha}}\right)g_{lm,k} - \frac{\partial\mathfrak{L}}{\partial g_{lm}}g_{lm,k}$$
$$= \frac{\partial}{\partial x_\alpha}\left(\frac{\partial\mathfrak{L}}{\partial g_{lm,\alpha}}g_{lm,k}\right) - \frac{\partial\mathfrak{L}}{\partial g_{lm,\alpha}}g_{lm,k,\alpha} - \frac{\partial\mathfrak{L}}{\partial g_{lm}}g_{lm,k}$$
$$= \frac{\partial}{\partial x_\alpha}\left(\frac{\partial\mathfrak{L}}{\partial g_{lm,\alpha}}g_{lm,k}\right) - \frac{\partial\mathfrak{L}}{\partial x_k}. \tag{11.19}$$

Thus the 'obstreperous' term is turned into a plain divergence. If we put

$$\mathfrak{t}^i_k = \frac{1}{2}\left(\delta^i_k\mathfrak{L} - \frac{\partial\mathfrak{L}}{\partial g_{lm,i}}g_{lm,k}\right), \tag{11.20}$$

(11.16) reads

$$(\mathfrak{T}^i_k + \mathfrak{t}^i_k)_{,i} = 0. \tag{11.21}$$

This goes to shew that the elementary ideas of 'density' and 'flux' apply, if at all, not to the components $\mathfrak{T}^i_k$ but to $\mathfrak{T}^i_k + \mathfrak{t}^i_k$. However, $\mathfrak{t}^i_k$ is *not* a tensor-density. But we will come back to this.

For computing its components explicitly (11.20) is not very convenient, because the partial derivatives of $\mathfrak{L}$ regarded as a function of the g_{ik} and $g_{ik,l}$ are not the most readily accessible. But we are not bound to these variables. We wish to use $\mathfrak{g}^{ik}$ and $\mathfrak{g}^{ik}{}_{,l}$ instead. Consider that

$$\delta\int\mathfrak{L}\,dx^4 = \int\frac{\delta\mathfrak{L}}{\delta g_{lm}}\delta g_{lm}\,dx^4,$$

but also

$$= \int\frac{\delta\mathfrak{L}}{\delta\mathfrak{g}^{lm}}\delta\mathfrak{g}^{lm}\,dx^4 = \int\frac{\delta\mathfrak{L}}{\delta\mathfrak{g}^{lm}}\frac{\partial\mathfrak{g}^{lm}}{\partial g_{rs}}\delta g_{rs}\,dx^4,$$

since the $\mathfrak{g}^{ik}$ are functions of the g_{ik}. It follows

$$\frac{\delta\mathfrak{L}}{\delta g_{lm}} = \frac{\delta\mathfrak{L}}{\delta\mathfrak{g}^{rs}}\frac{\partial\mathfrak{g}^{rs}}{\partial g_{lm}}$$

and

$$\mathfrak{T}^{lm}\frac{\partial g_{lm}}{\partial x_k} = -\frac{\delta\mathfrak{L}}{\delta\mathfrak{g}^{rs}}\frac{\partial\mathfrak{g}^{rs}}{\partial g_{lm}}\frac{\partial g_{lm}}{\partial x_k} = -\frac{\delta\mathfrak{L}}{\delta\mathfrak{g}^{rs}}\frac{\partial\mathfrak{g}^{rs}}{\partial x_k}.$$

By continuing exactly according to the pattern of (11.19) you obtain an alternative expression for our pseudotensor, viz.

$$\mathfrak{t}^i_k = \frac{1}{2}\left(\delta^i_k\mathfrak{L} - \frac{\partial\mathfrak{L}}{\partial\mathfrak{g}^{lm}{}_{,i}}\mathfrak{g}^{lm}{}_{,k}\right). \tag{11.20a}$$

To obtain the partial derivatives in the 'gothic, superscript' variables, form the complete differential of the identity (11.18)

$$\mathfrak{g}^{ik}{}_{,\alpha}d\Gamma^\alpha_{ik} - \mathfrak{g}^{ik}{}_{,k}d\Gamma^\alpha_{i\alpha} + \Gamma^\alpha_{ik}d(\mathfrak{g}^{ik}{}_{,\alpha}) - \Gamma^\alpha_{i\alpha}d(\mathfrak{g}^{ik}{}_{,k}) = -2d\mathfrak{L}$$

and replace the explicit derivatives in the first two terms (just as we did for establishing (11.18)) by those linear aggregates of the Γ's to which they are equal in virtue of the vanishing of $\mathfrak{g}^{ik}{}_{;l}$. Then those two terms turn out to be

$$= -\mathfrak{g}^{ik}\, d\mathfrak{L}_{ik} = -d\mathfrak{L} + \mathfrak{L}_{ik}\, d\mathfrak{g}^{ik}.$$

Thus $$-\Gamma^{\alpha}{}_{ik}\, d(\mathfrak{g}^{ik}{}_{,\alpha}) + \Gamma^{\alpha}{}_{i\alpha}\, d(\mathfrak{g}^{ik}{}_{,k}) - \mathfrak{L}_{ik}\, d\mathfrak{g}^{ik} = d\mathfrak{L}. \qquad (11.22)$$

This exhibits the partial derivatives we require; but since they must be symmetrical with regard to the superscripts, the second term has to be handled with care. It is best to rewrite it thus

$$\Gamma^{\alpha}{}_{i\alpha}\, d(\mathfrak{g}^{ik}{}_{,k}) = (\tfrac{1}{2}\delta^{\alpha}{}_{k}\, \Gamma^{\beta}{}_{i\beta} + \tfrac{1}{2}\delta^{\alpha}{}_{i}\, \Gamma^{\beta}{}_{k\beta})\, d(\mathfrak{g}^{ik}{}_{,\alpha}).$$

Hence $$\frac{\partial \mathfrak{L}}{\partial \mathfrak{g}^{ik}} = -\mathfrak{L}_{ik} = -(\Gamma^{\beta}{}_{\alpha k}\, \Gamma^{\alpha}{}_{i\beta} - \Gamma^{\beta}{}_{\alpha\beta}\, \Gamma^{\alpha}{}_{ik}),$$

$$\frac{\partial \mathfrak{L}}{\partial \mathfrak{g}^{ik}{}_{,\alpha}} = -\Gamma^{\alpha}{}_{ik} + \tfrac{1}{2}\delta^{\alpha}{}_{k}\, \Gamma^{\beta}{}_{i\beta} + \tfrac{1}{2}\delta^{\alpha}{}_{i}\, \Gamma^{\beta}{}_{k\beta}. \qquad (11.23)$$

These are the explicit expressions for the derivatives of which the second would be needed to make (11.20*a*) explicit.

For the sake of completeness let me insert a remark on the *homogeneity* of $\mathfrak{L}$, first with regard to the g_{ik} and $g_{ik,l}$. Of the latter *alone* it clearly is a quadratic form, from (11.17), since our Γ's are the Christoffel brackets. Of *both groups together* the Christoffel brackets are—since the g^{ik} are functions of degree -1 of the g_{ik}—homogeneous functions of degree zero. From (11.17) the same holds for the $\mathfrak{L}_{ik}$. Moreover, the $\mathfrak{g}^{ik}$ $(= \sqrt{-g}\, g^{ik})$ are obviously of degree $+1$ in the g_{ik}. Hence $\mathfrak{L}$ is homogeneous of degree $+1$ in the g_{ik} and $g_{ik,l}$ *together*, and therefore—since it is a quadratic form of the latter—it must be homogeneous of degree -1 in the former. All the statements of the last sentence remain true (on account of the one that precedes it), when g_{ik} and $g_{ik,l}$ are replaced by $\mathfrak{g}^{ik}$ and $\mathfrak{g}^{ik}{}_{,l}$. This entails, for example, the Euler-relations

$$\mathfrak{g}^{ik} \frac{\partial \mathfrak{L}}{\partial \mathfrak{g}^{ik}} = -\mathfrak{L},$$

$$\mathfrak{g}^{ik}{}_{,\alpha} \frac{\partial \mathfrak{L}}{\partial \mathfrak{g}^{ik}{}_{,\alpha}} = 2\mathfrak{L},$$

which can easily be verified by direct computation from (11.23) (you must use $\mathfrak{g}^{ik}{}_{;l} = 0$).

We proceed to discuss briefly the pseudotensor t^i_k and the conservation law (11.21). That the former is *not* a tensor has already been stated. It has the further deficiency that the array of components obtained from it by lowering the superscript with the help of the fundamental tensor is *not* symmetric (nor is, of course, the one you get by raising the subscript). Yet (11.21) holds in every frame provided that the pseudotensor is defined *in that frame* in the way we did define it. At first sight it is astonishing to find a relation holding in every frame though it does not treat of tensors only. But first we must not forget that ours is after all the tensor equation (11.15), only put into another form; secondly it exhibits *two* 'not properly covariant' features, namely in addition to the sham tensor a sham divergence, to wit the elementary one instead of the invariant one. These two features obviously compensate each other.

The components of the pseudotensor are sometimes spoken of as the gravitational energy-momentum-stress. In a way, they supersede the classical notion of gravitational potential energy, which has no other counterpart in Einstein's theory. They are not a very proper counterpart. It has, for example, been objected that in the field of an isolated mass-point you may make them vanish everywhere by a suitable choice of the frame. But, of course, for one *isolated* particle the notion of potential energy does not arise in classical theory either.

The most relevant aspect of the non-invariant shape (11.21) of the conservation laws is perhaps this: it throws particularly well into relief that in the present theory we must not expect conservation laws in the elementary sense to hold at large for $\mathfrak{T}^i_k$ in any frame. The change of the amount of T^4_k contained within a closed three-dimensional surface (i.e. the volume integral of $\mathfrak{T}^4_k$ extended over the interior) is *not* controlled by the flux $\mathfrak{T}^1_k$, $\mathfrak{T}^2_k$, $\mathfrak{T}^3_k$ through the boundary. There is a famous and singularly striking example of this, to wit: the total amount of T^4_4 (i.e. of energy or mass) contained in a closed expanding universe *decreases*. In simple models the loss can be computed and equals the amount of work the *pressure* would have to do to increase the volume, if a piston had to be pushed back as in the case of an adiabatically expanding volume of gas. Yet there is nothing like a piston nor any boundary at all through which energy could escape. To the pre-relativistic view (which might perfectly well endorse the idea of a closed expanding universe without surmising any connexion between the g_{ik} and the gravitational field) the energy is not lost, but stored as potential energy of the gravitating masses, which recede from each other.

Another example is that energy—and angular momentum—is 'carried away' through empty space ($\mathfrak{T}^i{}_k = 0$) from a system by the gravitational waves it emits when it has an inner motion such as to let its moments of inertia vary with time, say oscillate. The radiated energy need never turn up elsewhere as a T^4_4; but it can and will do so when those waves hit another system capable of partly absorbing them. The whole process is very similar to what we know so well from the classical theory of emission and absorption of electromagnetic radiation, except for the *true* energy tensor $T^i{}_k$ being zero inside the waves as long as they travel through empty space.

In such cases the exact conservation law for $\mathfrak{T}^i{}_k + \mathfrak{t}^i{}_k$ can serve to compute the energy-loss—or -gain, but to fix the ideas we shall speak of emission. The flux $\mathfrak{t}^1_4$, $\mathfrak{t}^2_4$, $\mathfrak{t}^3_4$ through a closed surface surrounding the system, but being itself situated in empty space, will give you the loss of $T^4_4 + t^4_4$ inside the surface. If the internal motions of the system are approximately periodical; if the frame is suitably chosen; and if the type of secular change due to the radiation can be qualitatively anticipated—as, for example, with a rotating rigid rod, which can clearly only change its angular velocity: then the *amount* of damping can be quantitatively inferred from that surface integral of the $\mathfrak{t}^k{}_4$-flux.

CHAPTER XII

GENERALIZATIONS OF EINSTEIN'S THEORY

AN ALTERNATIVE DERIVATION OF EINSTEIN'S FIELD EQUATIONS

The dynamical interaction of material bodies does not consist in their gravitational attraction alone. Electric and magnetic forces between them have been known for a very long time and have by Faraday and Maxwell been reduced to the notion of the electromagnetic field. In ordinary circumstances electromagnetic forces are, whenever they are observed at all, very much stronger than the gravitational pull, which is exceedingly weak unless at least one of the interacting pieces of matter is very large, of the size of a celestial body. Of late, one has been induced to admit that between the elementary particles (nucleons) which go to build up the nucleus of an atom there is a force (called the nuclear force) which is perceptible only at very small distances, but outweighs there even the strong electric repulsion between some of those particles. The field of this force is usually referred to as the meson field, for reasons on which we will not enter at the moment.

Ever since Einstein discovered his theory of the gravitational field in 1915, there have been unceasing attempts to generalize it so as to account in the same natural way for the electromagnetic field as well. Since the latter is in empty space described by an antisymmetric tensor of the second rank, the idea suggests itself at once that one should take the fundamental tensor g_{ik} to be non-symmetric, hoping that its skew part $\frac{1}{2}(g_{ik}-g_{ki})$ should have something to do with electromagnetism. But this plan meets with a certain *difficulty*. We had established Einstein's field equations in two steps. First we singled out the Christoffel-bracket affinity (9.9) cum (9.7) as the one that as it were 'naturally belongs' to the metric g_{ik}. This was done virtually by postulating the momentous identity (9.4), and deciding for a *symmetric* affinity. Only then could the Einstein tensor R_{ik} of this affinity be formed and the variational principle adopted,

$$\delta\int \mathfrak{g}^{ik}R_{ik}\,dx^4 = 0, \tag{12.1}$$

which was the second step, leading at once to Einstein's field equation for the vacuum, $R_{ik} = 0$, as explained on p. 98 to which we refer the reader.

The investigation in subsect. 1 of Chapter X goes to shew that even in the case of a metric properly speaking, I mean of a symmetric fundamental tensor g_{ik}, the Christoffel brackets quite obviously are not the one and only affinity which belongs to it in a natural way. However, this choice can at least be framed in the two postulates mentioned just before, or in words: a symmetric affinity which transfers the fundamental tensor into itself. *There is no obvious suggestion by what* (9.4) *should be replaced if g_{ik} is not symmetric.* This is the difficulty alluded to above, quite apart from the question whether in this case the symmetry postulate ought to be maintained for the Γ's or dropped. It might seem a very natural plan to admit also for non-symmetric g_{ik} the relation (9.4) as it stands. This has been tried but has failed. It is hardly worth while to shew it here *in extenso*. The failure will become understandable, when the truly natural generalization of (9.4) will emerge automatically, as it were.

The way out of this dilemma is shewn by a very important improvement on the derivation of Einstein's field equations. It is due to Palatini and avoids the 'first step': you have not to decide on the affinity beforehand, you get it together with the field equations from the variational principle at one go. It runs thus.

In (12.1) let $\mathfrak{g}^{ik}$ mean what it meant before, but R_{ik} the Einstein-tensor of an unspecified symmetric affinity $\Gamma^i{}_{kl}$. Take the g_{ik} *and* the $\Gamma^i{}_{kl}$ as the independent functions, to be varied arbitrarily, only retaining their respective symmetries on variation. You get anyhow

$$\int (\delta \mathfrak{g}^{ik} R_{ik} + \mathfrak{g}^{ik} \delta R_{ik})\, dx^4 = 0. \tag{12.2}$$

But now the two parts of this integral must vanish separately:

$$\int \delta \mathfrak{g}^{ik} R_{ik}\, dx^4 = 0, \tag{12.3}$$

$$\int \mathfrak{g}^{ik} \delta R_{ik}\, dx^4 = 0. \tag{12.4}$$

In the second line use (6.19), to wit

$$\delta R_{ik} = -(\delta \Gamma^\alpha{}_{ik})_{;\alpha} + (\delta \Gamma^\alpha{}_{i\alpha})_{;k}. \tag{12.5}$$

Integrating (12.4) by parts with respect to the semicolons (which is allowed since the affinity Γ to which the semicolons refer was assumed to be symmetric) you obtain

$$\int(\mathfrak{g}^{ik}{}_{;\alpha}\delta\Gamma^{\alpha}{}_{ik}-\mathfrak{g}^{ik}{}_{;k}\delta\Gamma^{\alpha}{}_{i\alpha})\,dx^4=0 \qquad (12.6)$$

or

$$\int(\mathfrak{g}^{ik}{}_{;\alpha}-\delta^k{}_\alpha\mathfrak{g}^{i\beta}{}_{;\beta})\,\delta\Gamma^{\alpha}{}_{ik}\,dx^4=0. \qquad (12.7)$$

The part of the bracket symmetric in i, k must vanish. That is easily seen to entail

$$\mathfrak{g}^{ik}{}_{;\alpha}=0. \qquad (12.8)$$

This is, of course, equivalent to (9.4) and entails (9.9), that is, our Γ's have to be the Christoffel brackets. Their Einstein-tensor is symmetric, and so (12.3) demands

$$R_{ik}=0, \qquad (12.9)$$

which, now that the Γ's are the brackets, are Einstein's equations for the vacuum.

THE EINSTEIN-STRAUS-THEORY

The singular merit of Palatini's derivation is that it can be extended straight away without ambiguity to a non-symmetric g_{ik}. The only decision to take in advance is, whether to uphold the symmetry demand for the *affinity* or to drop it. If you try to uphold it you fail. You get nothing new, only an absurdly arbitrary and useless supplement to the equations of pure gravitation. Again, we shall not trouble the reader by expounding this here, but go over to the case where the symmetry in both g_{ik} and $\Gamma^i{}_{kl}$ is dropped.

Apart from this we follow exactly the pattern of the preceding section, from (12.2) to the end. There is even a slight simplification inasmuch as the independent variations are now no longer restricted to symmetry, but entirely free; so we can now, for example, from (12.3) immediately infer (12.9), no matter whether or no R_{ik} may turn out to be always symmetric (of course it is not). On the other hand, there are two purely technical intricacies which require care. In the first place, for a non-symmetric affinity there is one further addition to (12.5), as indicated in (6.19). In the second place, the simple 'integration by parts with respect to the semicolon', by which

(12.6) was obtained from (12.4) cum (12.5), is equally restricted to symmetric affinities, for which it was proved in Chapter IV; in the non-symmetric case there is an additional term which has to be made out. Lest the conformity with the symmetric case be obscured by irrelevant details and the relevant results spread over too wide a space, we shall here once or twice give only the result of a transformation, relegating its technical complexities to an Appendix (p. 116), where the reader will find them if he thinks he needs them.

As regards the notation, we uphold the familiar relations between the four forms of the fundamental tensor, e.g.

$$\mathfrak{g}^{ik} = \sqrt{-g}\, g^{ik},$$

where g is again the determinant of the g_{ik} (not perhaps of their symmetric part). But, of course, in the relation defining the g^{ik} from the g_{ik}, viz.

$$g^{kl} g_{km} = g^{lk} g_{mk} = \delta^l_m, \tag{12.10}$$

the order of the indices is now relevant, and we adopt the afore-standing one. *No general scheme for raising and lowering indices is adopted now.* It is not needed and would be apt to cause confusion. We introduce the very convenient notation of Einstein and Straus, viz. we indicate the symmetric or skew-symmetric constituents of *anything* by underlining the couple of indices in question or putting a hook under them, respectively. For example,

$$\left.\begin{aligned} g_{\underline{ik}} &= \tfrac{1}{2}(g_{ik} + g_{ki}), \\ \Gamma^i_{\underset{\vee}{kl}} &= \tfrac{1}{2}(\Gamma^i_{kl} - \Gamma^i_{lk}). \end{aligned}\right\} \tag{12.11}$$

Now we turn to our variational equations (12.3) and (12.4). We have already pointed out that the first gives straight away (12.9), which we herewith register as the *first set of our field equations.* From (12.4) you obtain, by regular routine treatment† an unexpected result in lieu of (12.8). The simplest way of expressing it is in terms of *another* affinity, to be distinguished by an asterisk, and connected with the original one thus:

$$\left.\begin{aligned} {}^*\Gamma^i_{kl} &= \Gamma^i_{kl} + \tfrac{2}{3}\delta^i_k \Gamma_l, \\ \Gamma_l &= \tfrac{1}{2}(\Gamma^\sigma_{l\sigma} - \Gamma^\sigma_{\sigma l}). \end{aligned}\right\} \tag{12.12}$$

For the star-affinity $\quad {}^*\Gamma^\alpha_{k\alpha} = {}^*\Gamma^\alpha_{\alpha k},$ (12.13)

† See Appendix.

as you can verify at once. Then the result reads

$$\mathfrak{g}^{kl}{}_{,\alpha} + \mathfrak{g}^{\sigma l} {}^*\Gamma^{k}{}_{\sigma\alpha} + \mathfrak{g}^{k\sigma} {}^*\Gamma^{l}{}_{\alpha\sigma} - \tfrac{1}{2}\mathfrak{g}^{kl}({}^*\Gamma^{\sigma}{}_{\sigma\alpha} + {}^*\Gamma^{\sigma}{}_{\alpha\sigma}) = 0. \quad (12.14)$$

Our generalized field equations are (12.9) and (12.14) cum (12.12).

The spectacular event is (12.14). As was to be expected, it goes over into (12.8) in the symmetrical case, when the starred affinity coincides with the original one. But it is not the generalization of (12.8) anybody could have anticipated; for two reasons. *First*, and most momentously, the first member of (12.14) is *not* the invariant derivative of $\mathfrak{g}^{kl}$ with respect to the star affinity, for, as you see, *the order of subscripts is reversed in the third term.* It is a tensor density all right and you may call it a kind of invariant derivative, but not the ordinary one (see our comment on equation (3.7*a*) in Chapter III). *Only thanks to this reversed order do the equations* (12.14) *determine at least the starred affinity uniquely*, as in the symmetric case.† Otherwise they would not, and this was the principal cause of the failure, mentioned above, of the *naïve* generalization of (12.8).

The *second* unexpected feature is, of course, that the star affinity intervenes at all. It takes the rôle of *the* affine connexion in this theory, and it would be appropriate to drop the asterisk, if this were not apt to cause confusion. The *vector* Γ_l remains unspecified, it is *not* determined by the variational principle; but the star affinity is reduced by the injunction (12.13) from 64 to 60 independent components.

By contracting equation (12.14) once with respect to (l, α), then with respect to (k, α), and subtracting the resulting equations member by member, one gets

$$\mathfrak{g}^{k\alpha}_{\vee}{}_{,\alpha} + \tfrac{1}{2}\underline{\mathfrak{g}}^{k\sigma}({}^*\Gamma^{\alpha}{}_{\alpha\sigma} - {}^*\Gamma^{\alpha}{}_{\sigma\alpha}) = 0. \quad (12.15)$$

Hence, on account of (12.13)

$$\mathfrak{g}^{k\alpha}_{\vee}{}_{,\alpha} = 0. \quad (12.16)$$

Conversely (12.13) follows from (12.14) and (12.16), apart from singular cases. One may prefer to *replace* the injunction (12.13) by the set (12.16), interesting in itself since it has the general form of a Maxwellian set.

† See Appendix.

In much the same way as in the symmetrical case equations (12.14) can be turned into the equivalent but rather simpler form†

$$g_{kl,\alpha} - g_{\sigma l}{}^{*}\Gamma^{\sigma}{}_{k\alpha} - g_{k\sigma}{}^{*}\Gamma^{\sigma}{}_{\alpha l} = 0, \tag{12.17}$$

again with that peculiar order of subscripts in the last factor. Finally, by introducing the star affinity into (12.9) you easily get‡

$${}^{*}R_{kl} + \tfrac{2}{3}(\Gamma_{l,k} - \Gamma_{k,l}) = 0, \tag{12.18}$$

where the first term means the Einstein-tensor of the star affinity.

Equations (12.16)–(12.18) *can be regarded as the field equations of this theory*, equal in number to the number of unknown functions, to wit $64+4+16 = 84$. There is the following comment.

The 64 equations (12.17) are ordinary linear (non-differential) equations for the 64 components of the star affinity. They have a unique solution, corresponding to the Christoffel brackets of the symmetric case. It would mean a great reduction in the number of field equations if this solution could be written explicitly and inserted in (12.18). However, the attempt to follow this plan raises the suspicion that the explicit solution is much too complex for reaching a surveyable result in this manner.

A second, less relevant, remark is this. The equations (12.18) can be split into their symmetric and skew-symmetric parts:

$${}^{*}R_{\underline{kl}} = 0, \tag{12.18a}$$

$${}^{*}R_{\underset{\vee}{kl}} + \tfrac{2}{3}(\Gamma_{l,k} - \Gamma_{k,l}) = 0. \tag{12.18b}$$

The second set entails

$${}^{*}R_{\underset{\vee}{kl},i} + {}^{*}R_{\underset{\vee}{li},k} + {}^{*}R_{\underset{\vee}{ik},l} = 0 \tag{12.19}$$

and could be replaced by it, since from (12.19) our skew tensor must be the curl of *some* covariant vector, and Γ_l appears nowhere else in our field equations. However, this seems rather a step backwards. For, when faced with an equation of the type (12.19), you immediately infer that this tensor is a curl, and you write down something like (12.18*b*), which is simpler. The procedure is very familiar from one set of Maxwell's equations, which is precisely of this type.

A third remark concerns an interesting consequence of the set (12.17) alone. If you multiply it by g^{kl}, the first term is, according

† See Appendix. ‡ See Appendix.

to (12.10) the logarithmic derivative of g. The whole result may be written

$$\frac{\partial \log \sqrt{-g}}{\partial x_\alpha} = {}^*\Gamma^\sigma_{\underline{\underline{\sigma\alpha}}}, \qquad (12.20)$$

a relation very familiar from the symmetric case. We infer

$$\frac{\partial^* \Gamma^\sigma_{\underline{\underline{\sigma\alpha}}}}{\partial x_\beta} - \frac{\partial^* \Gamma^\sigma_{\underline{\underline{\sigma\beta}}}}{\partial x_\alpha} = 0. \qquad (12.21)$$

The underlining of the subscripts can be spared, if (12.13) or its equivalent (12.16) is taken into account. It is noteworthy that (12.21) holds, as a consequence of (12.17), in every frame, though the four quantities ${}^*\Gamma^\sigma_{\sigma\alpha}$ do *not* constitute a vector.

The further discussion is better postponed until the purely affine version of the theory has been expounded in the next section.

THE PURELY AFFINE THEORY

Can we not avoid introducing, with Palatini, *two* basic connexions of the space-time manifold, a quasi-metrical one by the g_{ik} and an affinity Γ^i_{kl}? Can one not go a step beyond Palatini and base a theory on affine connexion alone, which is after all the first and only one needed to obtain a basis for mathematical analysis (see Chapter III)?

But how are we to get an integrand for our variational principle? We can form the Einstein tensor of our affinity all right. But we cannot contract it with respect to its couple of covariant indices, to obtain a scalar density; nor have we any means of raising one of them.

A. S. Eddington pointed out as early as 1921 that the simplest invariant density you can build of the Einstein tensor alone is the square root of the determinant of its components (see our general remark Chapter II, p. 18). Both he and Einstein, in the following years, endeavoured to found on this basis a purely affine theory, unsuccessful for the reason, so I believe, that the affinity was from the outset taken to be symmetric (see Eddington's book *Mathematical Theory of Relativity*, whose later editions report comprehensively on Einstein's work as well). By dropping the symmetry demand one reaches a theory very similar to the one we outlined in the previous section, as we shall see forthwith.

So we take as our Lagrange function†

$$\mathfrak{H} = \frac{2}{\lambda}\sqrt{(-\operatorname{Det.} R_{ik})}, \tag{12.22}$$

and demand
$$\delta \int \mathfrak{H}\, dx^4 = \int \frac{\partial \mathfrak{H}}{\partial R_{ik}} \delta R_{ik}\, dx^4 = 0. \tag{12.23}$$

From general considerations—or immediately from the facts that the last integral is an invariant, and that δR_{ik} is an arbitrary tensor field—the array of partial derivatives constitute a contravariant density of the second rank. It is very convenient, if nothing more, to *call* this tensor density

$$\frac{\partial \mathfrak{H}}{\partial R_{ik}} = \mathfrak{g}^{ik} \tag{12.24}$$

and to supplement it by defining 'Latin' contra- and co-variant g-tensors in exactly the same way as before. For thus (12.23) comes to coincide in shape with (12.4), and all the consequences we drew in the preceding section from this 'second half' of the variational principle remain unchanged, we need not deduce them afresh. The 'first half', (12.3) and its consequence (12.9) are absent here. Instead, the set (12.24) posits a direct relation between the affinity and the 'metric' *in addition* to the one set by (12.14) or the equivalent (12.17). Formally the complete set of field equations of the present theory can be written down at once, in two ways: either we may insert, according to (12.24), the partial derivatives for $\mathfrak{g}^{ik}$ in (12.14) and (12.16), thus eliminating all the g's and leaving us with the components of the affinity as the only unknown functions; or we may insert into (12.24) the star affinity drawn from equations (12.17) and supplemented according to (12.12), in order to 'remove the asterisk'; in this case (12.16) would have to be joined as it stands and we would be left with the 'metric', i.e. with the g's, as the only unknown functions (virtually 16 in number). However, the second plan is practically ruled out by the complexity of the solution of (12.17), mentioned before.

Observe that in these considerations we have not yet drawn on the special form proposed for $\mathfrak{H}$ in (12.22). They would hold for *any* Lagrange function that depends on the R_{ik} alone. If this is not

† The constant λ and the minus sign are conventional and do not influence the result.

the case, they are modified. One might, for example, think of letting $\mathfrak{H}$ depend also on the 'second contraction' of the B-tensor, mentioned in Chapter IV, p. 51, and called S_{ik} there. In this case we should have

$$\delta \int \mathfrak{H}\, dx^4 = \int \frac{\partial \mathfrak{H}}{\partial R_{ik}} \delta R_{ik}\, dx^4 + \int \frac{\partial \mathfrak{H}}{\partial S_{ik}} \delta S_{ik}\, dx^4.$$

These two integrals would not have to vanish separately since the second depends ultimately on the same variation $\delta\Gamma^i_{kl}$ as the first. This, as I said, would modify and, indeed, complicate the field equations greatly, quite apart from the fact that there is no longer such an obvious suggestion what to take for $\mathfrak{H}$. We keep here to the choice taken at the outset.

I maintain that (12.24) is equivalent to

$$R_{ik} = \lambda g_{ik}. \tag{12.25}$$

Indeed, then

$$\mathfrak{H} = 2\lambda\sqrt{-g}, \tag{12.26}$$

and

$$\frac{\partial \mathfrak{H}}{\partial R_{ik}} = \frac{1}{\lambda}\frac{\partial \mathfrak{H}}{\partial g_{ik}} = -\frac{g g^{ik}}{\sqrt{-g}} = \mathfrak{g}^{ik}. \tag{12.27}$$

So even (12.9) suffers only a slight modification, it is superseded by (12.25). This is a very well-known set; in the symmetric case it goes over into what is called 'Einstein's equation with cosmological term'. We have not mentioned it in these lectures before. On any 'human' scale λ must be a very small constant, and the additional 'cosmological term' is practically irrelevant except in considerations concerning the structure of the universe. The purely affine theory is the only one that produces this term in a natural, non-premeditated way. It demands definitely $\lambda \neq 0$.

We can now, if we choose to, carry out our plan of eliminating the g-quantities altogether. Taking g_{ik} from (12.25) and inserting it into (12.17) we get the following self-contained set of field equations of this theory:

$$R_{kl,\alpha} - R_{\sigma l}{}^*\Gamma^\sigma{}_{k\alpha} - R_{k\sigma}{}^*\Gamma^\sigma{}_{\alpha l} = 0. \tag{12.28}$$

If we regard the starred components as abbreviations, explained by (12.12), our set contains only the original field-variables Γ^i_{kl} and nothing else. It is of the second order, as you would expect, and the 'irrelevant' constant λ has disappeared, as one must demand. Yet in the symmetric case it does embody Einstein's field equations *with* cosmological term! For a moment one might think that we have forgotten to include equations (12.16) in our set. But *they*—with the $\mathfrak{g}$'s replaced by their expressions in the R_{ik}—are consequences of (12.28), if our $^*\Gamma$'s are just short for (12.12).

A slightly different attitude is to regard the $^*\Gamma$'s with the injunction (12.13) and the vector Γ_l, as the unknown functions. Then one has (compare the transition from (12.9) to (12.18)) to regard the R_{kl} as abbreviations for

$$R_{kl} = {}^*R_{kl} + \tfrac{2}{3}(\Gamma_{l,k} - \Gamma_{k,l}). \qquad (12.29)$$

Again, and in the same sense as before, (12.16) is a *consequence.* It need not be included, unless for some reason or other one wished to avoid imposing (12.13) as an *a priori* injunction on the $^*\Gamma$'s.

The set (12.28) has no analogue in the Einstein-Straus-theory.

DISCUSSION OF THE PRECEDING THEORIES

For both theories—or versions—conservation identities (and a few others intimately connected with them) can be derived very much on the same lines as was done in Chapter XI for Einstein's theory. I will not deal with them here, but refer the reader to my paper in *Proc. R. Irish Acad.* **52**, A, p. 1, 1948.

For all that I know, *no special solution* has yet been found which suggests an application to anything that might interest us, save, of course, the well-known solutions in the symmetric case. It is known that even the latter are very limited in number. It is therefore not very astonishing that the much more intricate non-symmetric case should be obstreperous to the degree it is. Thus it is as yet undecided what interpretation of the various tensors and densities is most likely to let the theory meet observed facts. This holds not only for the skew-symmetric tensors, as for example, $\mathfrak{g}^{\overset{ik}{\vee}}$, etc., which, it is hoped, should have something to do with the electromagnetic field, and possibly with the nuclear field. We cannot even feel sure whether in the non-symmetric case the $g_{\underline{ik}}$ or the $\mathfrak{g}^{\underline{ik}}$ (or, less likely, the $\mathfrak{g}_{\underline{ik}}$ or the $g^{\underline{ik}}$) play the part of the corresponding tensorial entities describing the gravitational field in Einstein's theory. All four possibilities are distinctly different, but coincide, of course, in the limit, when all the tensors are symmetric. One must, so I believe, even be prepared to find that in general no quite clear-cut separation of the various fields exists; that they partly merge into one another in some manner that is difficult to foresee.

I do not think that the difficulty of finding exact solutions ought to deter us from thinking further about these theories. To my mind, they are the only ones to offer themselves as *natural* generalizations of an eminently successful predecessor. Suppose, for example, we were lucky and actually attained a lovely exact solution with cylindrical symmetry corresponding, in some admissible interpretation, to a localized electric charge surrounded also by a magnetic dipole field. What could we do with it? Could we consider it as a model of the spinning electron or proton? We could not. For we know that the classical interaction of such dainty little toys is altogether not competent to describe the actual electromagnetic interaction of the ultimate constituents of matter, and still less their interaction in the nucleus. In so far as any progress in the more complex features of this interaction (emission and absorption, particle creation and annihilation) has been made at all, it rests not on very complex classical solutions of the type alluded to just above, but on much simpler ones, to wit plane sinusoidal waves, which are just simple enough to be subjected to certain quantum-mechanical considerations. I suppose nobody deems this an ideal approach; but it will, so we hope, show us the way to better ones. This way is not likely to lead over very complicated 'particle-like' solutions. We may therefore, perhaps, console ourselves that such ones seem to be practically inaccessible in our case.

MATHEMATICAL APPENDIX TO CHAPTER XII

p. 109. We begin by extending to a *non-symmetric* affinity the rule for 'partial integration with respect to a semicolon', deduced in Chapter IV for a *symmetric* one. From the general expression for the invariant derivative of a contravariant vector density, to wit

$$\mathfrak{A}^k{}_{;i} = \mathfrak{A}^k{}_{,i} + \Gamma^k{}_{\sigma i}\mathfrak{A}^\sigma - \Gamma^\sigma{}_{\sigma i}\mathfrak{A}^k,$$

one obtains by contraction

$$\mathfrak{A}^k{}_{;k} = \mathfrak{A}^k{}_{,k} + 2\Gamma_k\mathfrak{A}^k.$$

(Γ_k is the abbreviation explained in (12.12), 2nd line.) This is the required modification of equation (4.3) in Chapter IV. If in the considerations that follow there, the supplementary term is taken into account, the general formula for partial integration is reached, viz.

$$\int (A^{\cdots}{}_{\cdots})(B^{\cdots}{}_{\cdots})_{;\alpha}\,dx^4 = \int [-(A^{\cdots}{}_{\cdots})_{;\alpha} + 2\Gamma_\alpha A^{\cdots}{}_{\cdots}]\,B^{\cdots}{}_{\cdots}\,dx^4,$$

where we have, however, suppressed a contribution from the boundary, since it will vanish in our case as in most applications.

After this preparatory movement we attend to (12.4), where we insert from (6.19)

$$\delta R_{ik} = -(\delta\Gamma^\alpha{}_{ik})_{;\alpha} + (\delta\Gamma^\alpha{}_{i\alpha})_{;k} + 2\Gamma^\alpha{}_{\underset{\vee}{\beta k}}\,\delta\Gamma^\beta{}_{i\alpha}.$$

By the ordinary routine of partial integration, using the rule we had just derived, we free the variations from the semicolons. The procedure is straightforward; but some deliberation is required to write the result in the simplest form. You will find no difficulty in verifying that

$$\int \mathfrak{g}^{ik}\delta R_{ik}\,d\tau = \int (\mathfrak{G}^{ik}{}_\alpha - \delta^k{}_\alpha \mathfrak{G}^{i\beta}{}_\beta)\,\delta\Gamma^\alpha{}_{ik}\,dx^4,$$

where $\quad \mathfrak{G}^{kl}{}_\alpha = \mathfrak{g}^{kl}{}_{;\alpha} - 2\mathfrak{g}^{kl}\Gamma_\alpha + \tfrac{2}{3}\delta^l{}_\alpha \mathfrak{g}^{k\beta}\Gamma_\beta + 2\mathfrak{g}^{k\beta}\Gamma^l{}_{\underset{\vee}{\alpha\beta}}.$

(This way of putting it is suggested by equation (12.7), to which everything must reduce, and does reduce, in the symmetric case.) Now you need only make the semicolon explicit to recognize that the *vanishing* of the last expression is precisely rendered by equation (12.14) cum (12.12) and (12.13).

p. 110. We prefer to deal with (12.17) and split it according to symmetry:

$$g_{\underline{kl},\alpha} - g_{\underline{\sigma l}}{}^*\Gamma^\sigma{}_{\underline{k\alpha}} - g_{\underline{k\sigma}}{}^*\Gamma^\sigma{}_{\underline{\alpha l}} - g_{\underset{\vee}{\sigma l}}{}^*\Gamma^\sigma{}_{\underset{\vee}{k\alpha}} - g_{\underset{\vee}{k\sigma}}{}^*\Gamma^\sigma{}_{\underset{\vee}{\alpha l}} = 0$$

$$g_{\underset{\vee}{kl},\alpha} - g_{\underset{\vee}{\sigma l}}{}^*\Gamma^\sigma{}_{\underline{k\alpha}} - g_{\underset{\vee}{k\sigma}}{}^*\Gamma^\sigma{}_{\underline{\alpha l}} - g_{\underline{\sigma l}}{}^*\Gamma^\sigma{}_{\underset{\vee}{k\alpha}} - g_{\underline{k\sigma}}{}^*\Gamma^\sigma{}_{\underset{\vee}{\alpha l}} = 0.$$

We must assume the determinant of the $g_{\underline{kl}}$ *not to vanish*, and we use them for index pulling. Take first the *symmetric case*, $g_{\underset{\vee}{kl}} = g_{\underset{\vee}{kl},\alpha} = 0$. Then from the second equation ${}^*\Gamma_{l\underset{\vee}{k\alpha}} = {}^*\Gamma_{k\underset{\vee}{l\alpha}}$, hence it vanishes. The first equation then gives ${}^*\Gamma^s{}_{\underline{kl}} = \left\{ {s \atop k\,l} \right\}$ by the procedure indicated at the bottom of p. 65. In the *general* (non-symmetric) *case* the *same* procedure, applied to the *first* equation, yields ${}^*\Gamma^s_{\underline{kl}}$ as a linear function of the ${}^*\Gamma^s_{\underset{\vee}{kl}}$, while applied to the *second* equation it gives ${}^*\Gamma^s_{\underset{\vee}{kl}}$ as a linear function of the ${}^*\Gamma^s_{\underline{kl}}$. By substituting alternately between the two you get for ${}^*\Gamma^s_{\underline{kl}} - \left\{ {s \atop k\ l} \right\}$ and ${}^*\Gamma^s_{\underset{\vee}{kl}}$ two series in ascending powers of $g_{\underset{\vee}{kl}}$ and its derivatives; for not too large values of these arguments these series are bound to converge. Since they represent rational functions, the solution could fail to exist only in exceptional cases.

p. 111. The relations (12.14) and (12.17) are equivalent without any prejudice, as (12.13), about the affinity. We have here to deduce the second set from the first. As in the familiar symmetrical case, it follows from (12.10) (which stipulates the g^{ik} as the normalized minors in the determinant g of the g_{ik}) that

$$\frac{\partial g}{\partial x_\alpha} = g g^{kl} \frac{\partial g_{kl}}{\partial x_\alpha} = -g g_{kl} \frac{\partial g^{kl}}{\partial x_\alpha}.$$

Thus

$$2\frac{\partial \lg \sqrt{-g}}{\partial x_\alpha} = -g_{kl}\frac{\partial g^{kl}}{\partial x_\alpha},$$

and also

$$g_{kl}\mathfrak{g}^{kl}{}_{,\alpha} = g_{kl}\left(\frac{\partial\sqrt{-g}}{\partial x_\alpha} g^{kl} + \sqrt{-g}\, g^{kl}{}_{,\alpha}\right)$$

$$= 4\frac{\partial\sqrt{-g}}{\partial x_\alpha} - 2\sqrt{-g}\frac{\partial \lg\sqrt{-g}}{\partial x_\alpha} = 2\sqrt{-g}\frac{\partial \lg\sqrt{-g}}{\partial x_\alpha}.$$

After this preparatory movement we attend to (12.14) and first multiply it by g_{kl}. The afore-standing equation gives us precisely the first term, and we easily get

$$2\frac{\partial \lg\sqrt{-g}}{\partial x_\alpha} - ({}^*\Gamma^\sigma{}_{\sigma\alpha} + {}^*\Gamma^\sigma{}_{\alpha\sigma}) = 0.$$

Next we return to (12.14) and multiply it by $g_{ks}g_{rl}$. We first obtain, always using (12.10),

$$g_{ks}g_{rl}\left(\frac{\partial\sqrt{-g}}{\partial x_\alpha} g^{kl} + \sqrt{-g}\, g^{kl}{}_{,\alpha}\right)$$

$$+\sqrt{-g}\,[g_{ks}{}^*\Gamma^k{}_{r\alpha} + g_{rl}{}^*\Gamma^l{}_{\alpha s} - \tfrac{1}{2}g_{rs}({}^*\Gamma^\sigma{}_{\sigma\alpha} + {}^*\Gamma^\sigma{}_{\alpha\sigma})] = 0.$$

The *first* term of *this* equation—according to the preceding one—amounts to

$$g_{ks}g_{rl}\frac{\partial\sqrt{-g}}{\partial x_\alpha} g^{kl} = \tfrac{1}{2}g_{rs}({}^*\Gamma^\sigma{}_{\sigma\alpha} + {}^*\Gamma^\sigma{}_{\alpha\sigma})$$

and thus cancels the *last* couple of terms. The *second* term, viz.

$$g_{ks}g_{rl}\sqrt{-g}\,g^{kl}{}_{,\alpha} = -\sqrt{-g}\,g_{rs,\alpha}.$$

It will be realized that we are thus left virtually with equation (12.17) which we set out to derive.

p. 115. We are to prove the relation (12.29) between the Einstein-tensors of the two affinities connected by (12.12)—the starred and the non-starred one. It is hardly necessary to print this proof. One uses the expression (6.17) of the Einstein-tensor and inserts the values of the non-starred Γ's in terms of the starred, drawn from (12.12). It so happens that the quadratic part of the Einstein-tensor is the same for the two affinities, since the additional terms all cancel. The result is (12.29).

p. 110. (*Note added on reprint in* 1954.) In the middle of that page we stated that the vector Γ_ι remains unspecified, it is not determined by the variational principle. May one perhaps simplify the theory by adding to the field equations, as obtained from the variational principle, the demand $\Gamma_\iota=0$? From (12.12) this would abolish the cumbersome distinction between the starred and the non-starred affinities, they would coincide. The question remained open for several years. It has recently been decided in the negative by A. Einstein and B. Kaufman (Volume in honour of Louis de Broglie, Paris, 1952, p. 321). The additional demand $\Gamma_\iota=0$ would imply (as these authors show by a careful and subtle investigation of *weak* fields) that the presence of a gravitational field, however weak, restricts the electromagnetic field, e.g. waves of light, in a perfectly inadmissible, nay, ludicrous fashion.

I wish to use this occasion for warning the reader, not to regard the interesting generalizations, briefly reviewed in Chapter XII, as anything like a well-established theory. It must be confessed that we have as yet no glimpse of how to represent electrodynamic interaction, say Coulomb's law. This is a serious desideratum. On the other hand we ought not to be disheartened by proofs, offered recently by L. Infeld, M. Ikeda and others, to the effect, that this theory cannot possibly account for the known facts about electrodynamic interaction. Some of these attempts are ingenious, but none of them is really conclusive.

Printed in the United States
By Bookmasters